Student Poster Sessions 2016

AA009193

Core Programming Area at the 2016 AIChE Annual Meeting

San Francisco, California, USA
13 - 18 November 2016

ISBN: 978-1-5108-3448-4

Printed from e-media with permission by:

Curran Associates, Inc.
57 Morehouse Lane
Red Hook, NY 12571

Some format issues inherent in the e-media version may also appear in this print version.

Printed by Curran Associates, Inc. (2017)

For permission requests, please contact AIChE
at the address below.

AIChE
120 Wall Street, FL 23
New York, NY 10005-4020

Phone: (800) 242-4363
Fax: (203) 775-5177

www.aiche.org

Additional copies of this publication are available from:

Curran Associates, Inc.
57 Morehouse Lane
Red Hook, NY 12571 USA
Phone: 845-758-0400
Fax: 845-758-2633
Email: curran@proceedings.com
Web: www.proceedings.com

Complete author information not provided in the original digital media.

TABLE OF CONTENTS

478047 Developing a Scalable System to Harvest Critical Materials from Industrial Waste with Gluconobacter Oxydans

Abstract:

Organic acid production by *Gluconobacter oxydans* has been proposed as part of a potential economical and environmentally friendly heterotrophic bioleaching process for recovery of rare earth elements from industrial waste. To provide organic acids in a quantity and at a rate useful for commercial bioleaching, continuous culture of the acid-producing organism will be necessary. The objective of this research project was to determine optimal conditions for continuous cultivation of *G. oxydans* for organic acid production for recovery of lanthanum and cerium from fluidized catalytic cracking catalyst. Cultures were subjected to dilution rates ranging from .05 hr^{-1} to .38 hr^{-1} to determine the optimum conditions for production of lixiviant agents conducive to leaching of inorganics. Previous studies have identified methods for maximizing production of gluconic acid and 2,5-diketogluconic acid from *G. oxydans*, yet the utilization of these and other unidentified *G. oxydans* metabolites for leaching of rare earth metals is largely unexplored. Continuous cultures exhibited sustained organic acid production for at least 110 hours, while batch fermentations showed slowed growth rates after 16 hours. Several parameters, namely optical density, dissolved oxygen, organic acid concentration and pH, were monitored to characterize the steady state conditions of the fermentation process. Leaching efficiency of organic acid solutions was quantified using inductively coupled plasma mass spectrometry. The observed trend was increasing acidity with decreasing dilution rates, as the lowest dilution rate of 0.05 hr^{-1} yielded the highest gluconic acid concentration of 176 mM and lowest pH of 2.2. Recovery of 41.5% to 47.9% of the total lanthanum and cerium in the fluidized catalytic cracking catalyst was attained with the organic acid solutions produced by *G. oxydans*. Leaching efficiency of the organic acid solutions increased with decreasing dilution rates, suggesting the effectiveness of lanthanum and cerium leaching from fluidized catalytic cracking catalyst is dependent on acidity of the lixiviant solution.

478306 Determination of Temperature Parameters for Optimization of a Plastic Pyrolysis Process for Fuel Oil

Abstract:

Municipal solid waste not properly disposed by major population centers of the developing world often finds itself dumped near slums or rural villages surrounding those regions. This waste is a serious environmental and health concern for the communities present in those regions. Lack of infrastructure, capital, and governmental policies continues the waste pile-up. Thus, methods of removing, recycling, and repurposing the otherwise pollutant waste are needed. In order to optimize the scale of municipal solid waste reduction, a process of plastic pyrolysis for the generation of a diesel and kerosene similar fuel oil has been proposed by the University of Kentucky Appropriate Technology and Sustainability (UKATS) research team. Several commonly occurring plastics such as High and Low Density Polyethylene, Polypropylene, and Polystyrene have been analyzed for their potential as hydrocarbon fuel oil. In order to optimize this chemical process, the temperature parameters around maximum yield and the removal of yield impeding, low viscosity hydrocarbon formation were to be determined. In order to achieve this, a small scale pyrolysis chamber was used to variably test constant temperatures. The small scale reaction chamber made of tubular stainless-steel that contained 75 grams of the desired plastic. The experimental results described the percent yields of plastic for variances in plastic type and the temperature of heating. This temperature scale began in a proposed range between 350°C and 450°C and was altered with reoccurring experimental findings. Along with data collection of individual plastics, a mixture of the LDPE, HDPE, PP and PS was also tested for appropriate applications. This data will then applied to the continuous heating of plastic inside the UKATS Processor. A modified rocket stove design, allowing the controlled heating through a wood fire at similarly high temperatures. The UKATS Processor can then be utilized in developing areas of the world offering an alternative recycling method with the potential to create jobs and alternative fuel.

Keywords: pyrolysis, sustainability

478797 A Sustainable Approach to Biodiesel Waste Minimization By Closing the Loop on Production

Abstract:

Sustainable energy is very current topic amongst many, including the 2016 presidential candidates. The idea behind sustainable energy is to be more environmental friendly, and to meet the needs of the present without compromising the ability of future generations to meet their needs. Over the past several years, a great deal of research has been done regarding biodiesel. Many processes have been developed in order to create the fuel, but each process creates various byproducts. In regards to the process utilizing used vegetable oil, the waste is glycerin. Glycerin can be purified and sold, but this is a very costly process and not very economical. Acrolein can be produced with glycerin as an intermediate. The first part of the process involves combining glycerin with phosphoric acid to yield sodium dihydrogen phosphate, which can be utilized as a catalyst for the production of acrolein from glycerin. Acrolein itself is a profitable compound, but it is also the first step in the production of acrylic acid, which is used in disposable diapers, emulsifiers, cosmetics, and pharmaceuticals. This process will essentially close the loop of the biodiesel process by using the byproduct to produce a more profitable compound and not produce any waste.

479670 Comparison of Cold Flow Properties Between Fatty Acid Methyl Esters and Fatty Acid Ethyl Esters

Abstract:

Biodiesel fuels are composed of the esters of fatty acid components of bio-based oils, such as canola oil, palm oil, poultry fat, other waste cooking oils, and many more. The alcohols used to convert these feedstocks into green fuels are typically methanol and ethanol, yielding fatty acid methyl esters (FAME) and fatty acid ethyl esters (FAEE), respectively.

Biodiesels have many characteristics that make them favorable in comparison to petrol-based fuels: renewability, carbon neutrality, lower sulfur emissions, biodegradability, and more. However, biodiesel fuels still have a few major shortfalls, including oxidative stability and cold flow properties, or operation at low temperatures. The latter shortcoming is the focus of this experiment.

Available literature on FAME biodiesel is much richer than FAEE biodiesel. FAEEs could potentially be quite valuable as biodiesel components, as ethanol can be renewably sourced, and pure FAEEs have lower cloud points than their corresponding FAME counterparts. Due to the shortage of FAEE data in literature, this experiment sought to establish a database of FAEE cold flow data, and draw comparisons between FAMEs and FAEEs.

480585 Use of Aerial NDVI Imagery in Determining Plant Water Status of Individual Grape Vines

Abstract:

With incidents of drought increasing in California and other areas around the world, increased water conservation in agriculture is more essential than ever before. As such, the developement of technology focused on understanding the plant water status of key agricultural crops is more important than ever before. In our study we sought to assess the feasibility of using aerial Normalized Difference Vegetative Index imagery to determine the plant water status of individual grape vines within a vineyard. Our research focused on fusing ground data with aerial data by comparing individual grapevine pruning weights to NDVI values extracted from aerial images. The ground and aerial data have different scales of resolution, both in space and time. We aimed to understand how to fuse these types of data. Additionally, our study focused on disambiguating spatial fluctuations into periodic parts (regularly spaced plants) and fluctuations due to irrigation and environment. We began our study by focusing on descriptive statistics (means and variances) and scatterplots (to understand any basic patterns). Next, we moved to more modern techniques, such as random forest and low-rank approximations to further understand patterns. In the conclusion of our study, we did not find NDVI adept in determining in time fluctuations in plant water strain of individual vines. However, our study revealed the potential for NDVI to be used in identification of live vegetation on a scale as fine as the individual grape vine.

478498 Effect of Zinc Telluride- Tungsten Oxide Composite for Water Splitting Using Cathodic Electrodeposition on Tin-Indium Substrate

Abstract:

The search for alternative sources of clean energy is even more apparent now versus twenty years ago due to the environmental awareness of todayâ€™s society. One source of energy which is explored is the photoelectrochemical splitting of water molecules. The development of the zinc telluride (ZnTe) sensitizers on the tungsten oxide catalyst (WO_3) which is electrodeposited on an Indium tin oxide substrate is a fairly new platform for the splitting of the water molecule. Our research involves a three step holistic method based on the principles of electrolysis where the indium oxide is initially coated with WO_3 and then with ZnTe. Tungsten oxide is selected for this process as it exhibits a 2.6 eV bandgap which makes it an effective source to harvest the solar spectrumâ€™s blue light as well as a high electrochromic capability. Whereas ZnTe is chosen to act as the electron collector for the cathode due to its negative conduction band edge of -1.63 eV, and it also displays a similar bandgap alignment of 2.26 eV. The objective of this work is to study the contribution of WO_3 decorated ZnTe in the splitting of H_2O. We will present the result of our investigation on cathodic electrodeposition of WO_3-ZnTe thin film on tin-indium substrate, with its catalytic properties as an electrode for H_2O splitting.

480659 Solar and UV Treatment of Human Waste to Yield NPK Plant Fertilizer

Abstract:

Various methods of removal of unwanted substances in human urine as applied to agriculture are examined. The current production processes of nitrogen and potassium for crop fertilizers are notably environmentally taxing. Phosphorus is primarily obtained through mining and will eventually run out. Human urine is abundant, renewable, and contains high concentrations of urea, phosphate, and potassium, all of which are excellent plant fertilizers. Human urine often contains pathogens, particularly if stored for long periods, as well as pharmaceuticals and their metabolites. This investigation covers various methods to degrade pathogens and large organic molecules in urine while maintaining high concentrations of fertilizer compounds. Methods include high temperature and pressure treatments, UV exposure, UV exposure with a TiO_2 catalyst, and absorption into activated carbon.

480592 Use of Lignin Biochar As a Catalyst Support

Abstract:

A higher demand for fuel and greener energy has driven the development of second generation biofuels. It has been proposed that liquid hydrocarbons can be synthesized from syngas using a carbon-rich biochar support impregnated with iron catalyst. Our main objective focused on synthesizing lignin-derived biochar support using a three-step process. The first step is producing lignin biocharâ€"lignin being a model component of biomassâ€"by hydrothermally carbonizing lignin at subcritical temperatures and pressures. The biochar was then activated by boiling it in a 0.1 M nitric acid solution reflux column. After drying, it was mixed with iron nitrate and left in an oven. To optimize the surface area and better the catalyst, the amount of impregnated time varied from one to seven days. Throughout the process, the surface area was measured using a BET surface area analyzer and the surface was examined with images from a Scanning Electron Microscope (SEM).

480729 Optimized Hydrothermal Liquefaction for High- and Low-Lipid Algae

Abstract:

Optimized Hydrothermal Liquefaction for High- and Low-Lipid Algae

Feng Cheng, Zheng Cui, Travis Le-Doux, Kwonit Mallick, Graham Hoffman, Jacqueline Jarvis, Neil Paz, Tanner Schaub, Nagamany Nirmalakhandan, Catherine E. Brewer*

Abstract

Third-generation biofuels produced from hydrothermal liquefaction (HTL) of algae have been developing rapidly. Some advantages of algae-derived oils via hydrothermal liquefaction are attributed to high algae growth rates, strong CO_2-mitigation potential, and avoidance of feedstock drying requirements. In HTL, subcritical water (270-370°C and 60-210bar) depolymerizes lipids, proteins and carbohydrates in algae, leading to high yield of bio-crude oil, which can be upgraded to transportation fuels. In this study, we delineate biocrude composition as influenced by operating condition change for temperatures of 310-350°C, residence time of 5-60min, and solid algae content of 5-10wt.%) for both the high-lipid microalgae *Nannochloropsis salina* and the low-lipid microalgae *Galdieria sulphuraria.* The bio-crudes are characterized by Fourier transform ion cyclotron resonance mass spectroscopy (FT-ICR MS), fatty acid methyl ester (FAME) analysis by gas chromatography mass spectroscopy (GC/MS), oxy-combustion calorimetry, and elemental analysis (CHNS). These results are used to optimize species-specific operating conditions for a 1.8 L batch reactor and to predict optimized conditions for new algae strains based on their feedstock compositions and knowledge of compound degradation pathways. In addition, bio-crude oil characterization results will be used to select initial conditions for a pilot-scale continuous flow HTL reactor, designed to produce char-free biocrude oil from low-solid-content feedstock without centrifugation.

480706 Exploiting Natural Processes to Reduce the Need for Petroleum-Based Fertilizers: The Role of Bacterial Chemotaxis in Nitrogen Fixation of Leguminous Plants

Abstract:

Biological nitrogen fixation is a natural process that supplies nitrogen to the soil and reduces the need for petroleum-based fertilizer. Soil bacteria belonging to the genus *Rhizobium* form a symbiotic relationship with leguminous plants to fix atmospheric nitrogen into ammonia, a form that can be taken up through the vascular system of all plants once released into the soil. This process operates through a specific signaling cycle that induces the migration of rhizobia and formation of nitrogen-fixing nodules on the roots of the host legume.[2] It has been shown experimentally that the directed migration of the bacteria is governed by a mechanism called chemotaxis.[3,4] Chemotaxis is a physical motile response of bacteria to environmental signals and concentration gradients. It occurs when chemical cues bind with receptor molecules of the rhizobia and alter the swimming pattern and bias the direction of movement towards the source. This chemotactic behavior of the rhizobia is an important trait that facilitates the initial contact and adsorption of symbiotic rhizobia to the host root surface, increases the efficiency of nodule initiation, and increases the rate of infection development.[4,5]

The aim of this study is to quantify the effect that soil moisture content has on rhizobia chemotaxis and nodule formation. A series of rhizotron experiments were conducted to demonstrate that moisture content impacts nodule formation, with special attention given to nodule primordia formation. *Vigna unguiculata, commonly known as the cowpea, was grown in hydroponic conditions on horizontally oriented, rockwool layered petri dishes. Bacteria (Bradyrhizobium spp) were inoculated via a peat slurry at the end opposite the seed, as shown in figure 1. This rhizotron design provided a 144 cm^2 observation area for root growth, bacterial migration, and nodule formation. Observation of nodule primordia formation at root locations farthest from the inoculation site indicated chemotactic migration had occurred. By systematically reducing the amount of water supplied, a water content level was reached, at which there was an absence of nodule formation. This suggested a percolation threshold at which a continuous liquid pathway for bacterial motility no longer existed. The next step of this project is to quantify the impact of moisture content. This will be done using a Buchner-funnel tensiometer. With this device, the water content can be controlled in certain pore sizes by adjusting the hydrostatic pressure. With this experiment, it will be possible to identify the desired levels of moisture in soil to facilitate nodule formation. These results will be compared to climate conditions in order to assess the potential for optimizing this natural process for a more sustainable alternative to petroleum-based fertilizers for crop nutrition.*

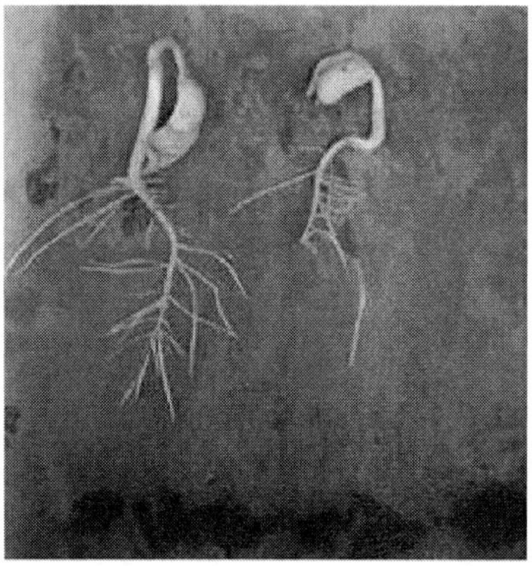

Figure 1: Example of rhizotron set-up from above view point. Chemotactic migration can be determined from location of nodule formation.

1. Food and Agricultural Organization (2008). Retrieved November 26, 2015, from http://www.fao.org/soils-portal/soil-degradation-restoration/en/

2. Bergman, K. (1990, January). The Role of Bacterial Chemotaxis in Initiation of Rhizobium Legume Symbiosis for Nitrogen-Fixation. *Journal of Chemical Ecology, 16(1),* 114-115.

3. Caetano-Anollés, G., Wall, L. G., De Micheli, A. T., Macchi, E. M., Bauer, W. D., & Favelukes, G. (1988). Role of motility and chemotaxis in efficiency of nodulation by Rhizobium meliloti. *Plant Physiology, 86*(4), 1228-1235.

4. Soby, S., & Bergman, K. (1983). Motility and chemotaxis of Rhizobium meliloti in soil. *Applied and environmental microbiology, 46*(5), 995-998.

480762 Resolving the Kinetic Parameters of Cellulose Pyrolysis

Abstract:

Biomass pyrolysis is an emerging technology used to convert renewable resources into energy, biofuels and chemicals. The gaseous, liquid, and char products are alternatives to conventional resources and contribute less greenhouse gases to the environment. Thus, there has been great interest in research on the kinetic mechanisms governing biomass pyrolysis, much having been performed using thermal gravimetric analysis (TGA). Nonetheless, the actual kinetic parameters for biomass pyrolysis, such as activation energies, are still vigorously debated. Among the most contentious issues is heating rate effects and activation energy invariance as a function of the same. Our hypothesis is, in fact, the opposite; the kinetic parameters for cellulose pyrolysis are a function of heating rate. To test this hypothesis, cellulose pyrolysis data was collected from four TGA experiments run at heating rates of 1, 10, 50, and 150 K/min. Three methods were then utilized to extract kinetic parameters: (1) linear regression of single heating rates, (2) co-regression of multiple heating rates, and (3) application of an isoconversional kinetic analysis. The collective outcomes suggests that activation energy appears to depend on the heating rate since the activation energy for higher heating rates is significantly different from those of lower heating rates. Due to this finding, co-regression methods could not produce reliable or accurate results. Surprisingly, isoconversional methods produce parameters that are inconsistent with single heating rate analysis, and remains a yet unresolved issue.

480783 Physicochemical Characterizations of Distillate Fractions of Biocrude Oil Converted from Wet Biowaste Via Hydrothermal Liquefaction

Abstract:

Physicochemical Characterizations of Distillate Fractions of Biocrude Oil Converted from Wet Biowaste via Hydrothermal Liquefaction

Zhenwei Wu[a], Wan-Ting Chen[a], Yuanhui Zhang[a*], Lance Schideman[b], B.K. Sharma[b]

[a] Department of Agricultural & Biological Engineering, University of Illinois at Urbana-Champaign, Urbana, 61801

[b] Illinois Sustainability Technology Center, Champaign, Illinois, 61820

The goal of this study is to understand the physicochemical properties of distillate fractions of biocrude oil converted from wet biowaste, such as swine manure and food processing waste, via hydrothermal liquefaction (HTL). HTL is a technology used to reduce complex organic materials into a biocrude oil fraction, an aqueous fraction, a gaseous fraction, and a solid residue fraction. Our previous study showed that biocrude oil converted from wet biowaste via HTL contains comparable and higher heating values (HHV) to transportation fuels such as gasoline. However, the application of HTL biocrude oil still remains unknown because up-scaled production and proper separation of HTL biocrude oil are still underdeveloped. As a consequence, this study aims to investigate proper separation methods for HTL biocrude oil. It is hypothesized that distillation can efficiently separate HTL biocrude oil into different fractions because preliminary study shows that the oxygen and nitrogen content in the distillates can be reduced to 2-5% while the heating values can be increased up to 41-47 MJ/kg. Biocrude oil converted from swine manure and food processing waste will be separated by distillation into 8-10 fractions in this study. Detailed physicochemical characterizations, such as GC-MS, FTIR, and TGA analyses, will be conducted with the distillates to investigate their properties for transportation fuel application. Ultimately, it is envisaged that this study can generate knowledge for understanding the fuel properties as well as provide insights for future catalytic hydroprocessing of HTL biocrude oil.

477681 Unraveling the Interactions of Homogenous Lewis Acid Catalyst (aqueous Tin(IV) Chloride) Species with Glucose By Means of Raman Spectroscopy

Abstract:

: Utilizing renewable resources, such as lignocellulosic biomass, for the production of fuels and chemicals has attracted significant interest in recent years. A characteristic example involves the conversion of cellulose to 5-Hydroxymethylfurfural (HMF) through a series of catalytic steps. A bottleneck in this reaction network is the conversion of glucose to fructose via isomerization reaction. Lewis acid catalysts, such as Sn-Beta, are essential for the selective isomerization of glucose to fructose. In this work, we utilize Raman spectroscopy to study the interactions between aqueous tin chloride species (homogeneous Lewis acid catalyst) with glucose in an effort to establish structure function-reactivity relationships for the glucose isomerization reaction. Specifically, spectroscopic characterization of $[Sn(H_2O)_6]^{4+}$, $[SnCl(H_2O)_5]^{3+}$, $[SnCl_2(H_2O)_4]^{2+}$, $[SnCl_3(H_2O)_3]^{+}$, $[SnCl_4(H_2O)_2]$, & $[Sn(OH)_6]^{2-}$ complexes in aqueous solutions was performed by deconvoluting pertinent spectral envelopes (50-250cm^{-1}, 290-350cm^{-1} and 400-700cm^{-1}). The deconvolution of the binary Tin(IV) Chloride/water as well as the ternary Tin(IV) Chloride/glucose/water solutions has revealed significant changes in the vibrational modes of specific aqueous tin species. A sharp decrease in the 65 cm^{-1} band of Tin(IV) Chloride with the addition of glucose in the aqueous solution, could possibly be attributed to the binding of asymmetric tin species with glucose molecules. To further characterize the complexation between glucose and aqueous tin species, isotopic exchange studies using deuterated water will be discussed. Additionally, to determine the exact symmetry behind these dissociated Tin(IV) Chloride species, identifying the respective depolarization ratio for each significant Raman band has also been achieved. We also couple Raman spectroscopic measurements with chemical equilibrium calculations using Visual MINTEQ in order to provide information about changes in the species equilibrium as well as potential interactions between tin species

and glucose. For the first time, the full characterization of aqueous Tin(IV) Chloride from 50-700cm^{-1} along with the specific vibrational modes for each Raman active band have been elucidated.

477589 Theoretical Investigation of the Decomposition of Glycerol over Pt(111)

Abstract:

Glycerol is a significant byproduct in biodiesel production and identifying efficient technologies for its conversion into higher value products is highly desirable. Due to its high degree of functionality, glycerol has the potential to be catalytically converted into a number of value-added chemicals such as ethylene glycol, 1,2-propanediol, 1,3-propanediol, and lactic acid, as well as syngas. Gaining a firm understanding of the underlying mechanism involved with glycerolâ€™s conversion over metal surfaces is complicated because of the large number of possible reaction paths. It is for this reason that (computational) research tends to favor the investigation of smaller molecules with lower functionality. Nevertheless, availability of computational resources has increased sufficiently such that trends in the hydrogenolysis of glycerol can be studied from first principles over selected metal surfaces. Pt(111) has been shown to favor C-C bond cleavage over C-O cleavage after high levels of dehydrogenation with the assistance of an empirical scaling scheme to evaluate the binding energies of intermediates. Here, I present Bronsted-Evans-Polanyi (BEP) relations for glycerol over a Pt(111) surface evaluated for a larger number of empirical results in order to gain a deeper understanding of the favorability of dehydrogenation, deoxygenation and decarbonylation relative to each other. All elementary rate constants in the model have been determined from density functional theory (PBE+D3) and harmonic transition state theory.

477830 Theoretical Investigation of the Degradation Pathway of Phosphorus Containing Nerve Agents on Metal Oxides

Abstract:

The mechanism for the decomposition of chemical warfare agents (CWA), particularly those classified as nerve agents, is investigated. It is experimentally known that metal oxides are capable of decomposing nerve agents with enhanced catalytic activity occurring at vacancy sites. To date, however, a detailed mechanistic study revealing the path of degradation is lacking. The focus of this project is to investigate the different surface planes of zinc oxide as a catalyst for hydrolysis of phosphorus-containing agents and their simulants. In this study, the molecules POH2, POH3, and POH2F, whose phosphorus containing functional group PO is incorporated in most nerve agents, is used as a starting point of interaction. In order to reveal the bonding nature with zinc oxide many different adsorption sites were chosen on both 002 and 100 surface planes. In addition, vacancy sites in the zinc oxide cuts were considered in the calculations. Binding energies were obtained by employing density functional theory (DFT), using the commercially available software VASP. In parallel, experiments were performed to produce zinc oxide powders rich in vacancies by first treating zinc oxide with hydrogen peroxide to form zinc peroxide followed by annealing at which point molecular oxygen is desorbed from the surface, leaving behind a vacancy. These samples were characterized through various techniques and catalytic activity was tested against CWA simulants. Through this combined computational and experimental approach, it is predicted that the results obtained here will not only provide a mechanistic insight towards degradation, but aid in the future development of more efficient catalyst against CWAâ€™s.

477683 Loop Reactors - Revolutionizing Reaction Engineering

Abstract:

Loop Reactors have set a benchmark when it comes to reactions that have important consideration of uniformly controlled mixing and intense dispersion without mechanical agitation. Apart from being easily operable, these reactors have a very peculiar advantage of having a simple construction and requirement of very less power for operation. Hence loop reactors are gaining wider acceptance in the chemical industry. Numerous aspects of the loop reactors have been studied and documented in the past - gas hold up studies, mass transfer studies, flow distribution studies and residence time distribution studies, being the most researched ones.

After a systematic literature review, it was found that computational fluid dynamic approach is the new addition to research domain of the loop reactors. Modelling and simulation using various reaction models and turbulence models are a new trend among researchers and various papers have been published on the same. Recently, process control studies have also been explored briefly in context of loop reactors, however this field needs more attention. Thermal hydraulics study on loop reactors is a vast field with numerous aspects and extensive research is being performed on the same. Nuclear research and techniques are focused on thermal hydraulics to a large extent and study of the same in loop reactors has immense scope. Present work is a critical analysis of various design factors which elucidate multiple aspects of loop reactors.

478128 O-Nitrophenyl Cellobioside As an Active Site Probe for Family 7 Cellobiohydrolases

Abstract:

Interactions between *o*-nitrophenyl-Î²-D-cellobioside (*o*NPC) and two cellobiohydrolases, Cel7A from *Hypocrea jecorina (Trichoderma reesei)* and Cel7D from *Phanerochaete chrysosporium* were studied by fluorescence spectroscopy and kinetic tests. The hydrolysis rate is slow enough that the substance may be treated as nonreactive in equilibrium binding studies. *o*NPC quenches the natural fluorescence of Cel7A and Cel7D by radiationless energy transfer. Addition of cellobiose recovers the fluorescence of Cel7A, indicating that *o*NPC binds preferentially to the +1,+2 subsites of the enzyme, to which cellobiose is known to have highest affinity. *o*NPC was used as indicator ligand to determine the association constants of cellobiose for catalytically inactive mutants of Cel7A by displacement binding experiment. The experimental data are supplemented by structural modelling of *o*NPC into the active site of the enzymes studied. Further, molecular dynamics simulations were performed to identify active site contributions to affinity, and free energy calculations (molecular mechanics Poisson-Boltzmann surface area) were conducted to understand preferential subsite binding. Kinetic studies have shown that oNPC is far less susceptible to hydrolysis by Cel7A than other active site probes. Molecular dynamics simulation has shown a wealth of hydrogen bonds formed and highly favorable interaction energy when oNPC is bound in the +1,+2 subsites.

478175 Robust Catalyst for the Dehydrogenation of Formic Acid

Abstract:

Robust Catalyst for the Dehydrogenation of Neat Formic Acid

Elyse Kedzie

According to the U.S. Energy Information Administration (EIA), global energy consumption is projected to grow by 48% by year 2040.[1] This presents a clear obstacle for the worldÕs continued economic growth. The rate of energy consumption is much greater than its production; thus, it is necessary to find new and methodical ways to harvest energy and convert it to portable fuels. In regards to transportation, liquid petroleum is most commonly used to operate our vehicles, which creates sustainability and global carbon balance issues. A developing form of alternative energy is the use of hydrogen. However, because it exists as a gas at ambient temperatures, it has low energy density and some safety risks.

Formic acid (FA) is a promising hydrogen storage material because it is an inexpensive, nontoxic, easily handled liquid.[2] FA can be converted to CO_2 and H_2, the latter being a clean energy carrier with application in fuel cells in which the byproduct is water. In this research, we investigated iridium complex **1** and showed it to be the first efficient homogeneous catalyst for the decomposition of *neat* formic acid (FA) into H_2 and CO_2.[3] Complex **1** works under mild conditions in the presence of air, is remarkably robust with a turnover number (TON) greater than 2,000,000 (which is more than twice the highest TON previously reported), and is reusable without regeneration. Furthermore, the decomposition reaction proceeds with high selectivity (carbon monoxide, CO, a possible byproduct which poisons fuel cell catalysts) is not produced at a detectable level.

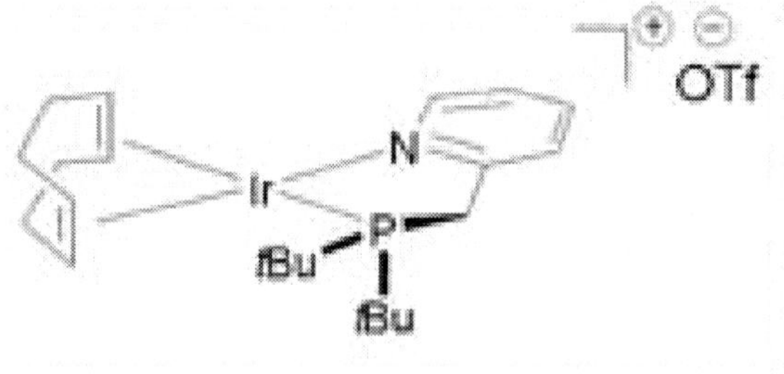

Figure 1: Iridium Complex 1

This is the first truly practicable, high throughput system for generating H_2 from formic acid for fuel cell applications.

1. ÒEIA projects 48% increase in world energy consumption by 2040Ó. *US Energy Information Administration: Independent Statistics and Analysis*. 2016.
2. Yamamoto, Y. *Chem. Rev.* **2012**, *112*, 4736-4769.
3. Celaje, J. J. A.; Lu, Z.; Kedzie, E. A.; Terrile, N. J.; Lo, J. N.; Williams, T. J. ÒA Prolific Catalyst for Dehydrogenation of Neat Formic AcidÓ *Nature Commun.* **2016**, 7, 11308

478229 Nickel and Iron Catalysts Supported on Cellulose-Based Carbon Aerogels: An Application of Bcfp

Abstract:

Nickel and iron catalysts for biomass pyrolysis were prepared on carbon aerogel (CAG) supports made from cellulose nanofibers. The nanofibers were impregnated with $(NH_4)_2SO_4$ and subjected to varying heating rates (10 or 20 °C / min), maximum temperatures (900, 1000, or 1100 °C), and residence times (0, 1, 2, or 5 h) under nitrogen gas to yield carbon aerogels (>90% carbon). The CAG featured high thermal resistance, average pore radius of 2-3 nm, and surface areas of up to 809 m^2 / g; surface areas were much lower for CAG produced at higher temperatures and longer residence times. CAG were also characterized using X-Ray diffraction (XRD) and Fourier Transform Infrared Spectrometry (FTIR). CAG 90 is an abbreviated notation for CAG produced at 900 °C with 0 h residence time and 10 °C / min heating rate. Samples of CAG 90 were impregnated with nickel nitrate or iron acetylacetonate via incipient wetness. They were then dried and reduced to produce 10 wt% nickel and iron catalysts. Biomass catalyzed fast pyrolysis was performed with Eucalyptus globulus and catalyst in a 1:5 mass ratio using a micro pyroprobe coupled to gas chromatography mass spectrometer (py-GCMS). Both catalysts reduced the selectivity for carboxylic acids, furans, and phenols while increasing the selectivity for larger aromatic compounds. The pyrolysis temperature of 500 °C and duration of 6s failed to yield hydrogen, methane, or other non-CO2 gases.

478254 Self-Supporting Monoliths from Hollow Core-Shell Ni@SiO2 Nanostructures

Abstract:

Metal nanoparticles are finding widespread application in various industries due to their unique properties compared to their bulk counterparts. However, beyond advantageous properties, metal NPs also show a greater tendency to form larger clusters, mitigating their "nano-effects". To counter this tendency, encapsulation of metal NPs in stable oxide shells, such as silica shells, has been shown previously to increase the thermal stability of the embedded metal NP due to reduced ability for Oswald Ripening and coalescence. While these core-shell configurations hence show great promise for robust application of NPs, these materials are still in nano-powder form and constitute significant health and safety concerns. Furthermore, nano-powders pose an engineering challenge due to excessive pressure drop when used in a catalytic bed, for example. While these issues could be overcome by depositing these core-shell structures onto conventional supports, a self-supporting monolithic structure made from such core-shell materials would be a uniquely efficient and potentially flexible configuration.

In the present work, we present results from an investigation into forming such self-supporting monolithic structures from hollow Ni@SiO2 core-shell catalysts. The aim of the study is the formation of monoliths from Ni@SiO2 without the aid of sacrificial chemicals (such as binders) or support structures, and, ultimately, to control the meso- and macroporosity of the resulting monolith while maintaining a minimum mechanical stability. Towards this goal, several different approaches were tested based on a cross-linking of the oxide shells with and without additional silica reinforcement. The resulting monoliths were characterized via SEM and TEM and tested for their rupture strength. Based on the results, we developed an optimized protocol to form stable, self-supporting monoliths from Ni@SiO2 core-shell materials. Since the procedure is based on cross-linking the silica matrix, it is expected to be transferrable to other hollow and non-hollow metal@SiO2 systems. We are currently evaluating these structures in catalytic experiments, comparing the monolith against a conventional catalytic fixed-bed.

478326 Preparing Supported Metal Catalysts Coated with Thiols and Phosphonates

Abstract:

In the current study, dually coated supported metal catalysts are prepared; thiols are used to modify metals, with the sulfur head group binding to the surface, while phosphonic acids are used to modify the support. Both thiol and phosphonic acid modifiers form self-assembled monolayers(SAM) on the catalyst surface. Cinnamaldehyde hydrogenation was used to probe the performance of the as-prepared catalysts; the selectivity to cinnamyl alcohol could be potentially improved without significant decreases in activity by coating two self-assembled monolayers on the surface of the catalyst. For the dually coated catalyst, the thiol modifier mainly increased selectivity as was expected from previous work, while the phosphonate coating mainly increased the activity. When adding both thiols and phosphonates to the catalysts in different orders, it was found that the selectivity resembled the second SAM that was added to the catalyst. These changes in selectivity were less drastic than the selectivity changes from a single SAM on the surface. For the activity of the catalyst, all of the catalysts with phosphonate coating showed increased activity for cinnamyl alcohol production in comparison to the uncoated Pt/Al_2O_3. The largest increase was seen when phosphonate was added after the thiol, likely from the thiol opening the selective active sites that favor cinnamyl alcohol production while the larger amount of phosphonate effected the electronic properties on the surface to increase the rate.

478352 Effect of Pore Structure on CO2 Adsorption Characteristics of Aminopolymer Impregnated MCM-36

Abstract:

The adsorption properties of a 2-dimensional, pillared, porous silicate material when loaded with an amine containing polymer were investigated. In this study, MCM-36, a stacked lamellar silicate, was used as a support for loading with different amounts of polyethylenimine (PEI) to augment the base materialâ€™s adsorptive properties. Based on the results of this research, high amine loading led to blockage of the mesoporous channels within the MCM-36 and degraded carbon dioxide diffusion into the solid and thus its adsorption. The upper limit amine loading placed upon inward carbon dioxide diffusion implied that there is an ideal amount of amine loading for any particular porous support that can improve the selectivity and capture of carbon dioxide compared to that of the starting material. This finding can be extrapolated to test the capture properties of other amine loaded supports in order to quickly determine their usefulness in future applications. Additionally, these results suggest that a different support structure that can accommodate amine loading without hindering diffusion rates may be necessary for effective carbon dioxide capture with solid adsorbents.

478605 Catalytic Analysis of Copper Porphyrine in the Conversion of Methane to Methanol through Computer Simulations

Abstract:

In todayâ€™s world the demand for methanol is at a high. Current production of methanol requires a large amount of energy, which in turn can be expensive and dangerous. In an attempt to optimize the process, research has been conducted on the use of a catalyst that can reduce the energy required to complete the conversion from methane to methanol. This work includes the analysis of oxygen-centered copper porphyrine (O-Cu/P), used as a catalyst placed on functionalized sheet of graphene. The system was evaluated via density functional theory and nudged elastic band theory which assisted with developing an understanding for the reaction energetics, and reaction mechanisms when using this catalyst. Values obtained were able to validate that O-Cu/P created a thermodynamically favorable reaction with an activation energy of 0.29 kJ/mol and 1.33 kJ/mol for the steps of the C-H bond scission and the formation of methanol. The main advantage of this work is the recognition that metal-oxo active center of O-Cu/P can reduce the energy required in the current production of methanol.

478637 High Pressure High Temperature Annealed TiO_2 for Photocatalytic Biomass Reforming

Abstract:

Hydrogen gas is considered a form of green energy due to its clean emissions. However, about 95% is produced from unsustainable processes like methane steam reforming and coal gasification. Photocatalytic generation of hydrogen from renewable feedstocks such as biomass or landfill-derived methane will increase the sustainability of hydrogen gas production. This project focuses on the synthesis, characterization, and photocatalytic activity of annealed Anatase phase TiO_2 nanoparticles. TiO_2 is stable under catalytic conditions and has high photocatalytic activity. However, its large bandgap of 3.2 eV corresponds to activity under UV irradiation which comprises about 5% of incident solar radiation on Earth. We reduce this large bandgap to extend to the visible light range by inducing electronic and geometric defects in the TiO_2 structure by annealing Anatase under Hydrogen in high pressure and temperature conditions. This contributes to the formation of oxygen vacancies and Ti^{3+} centers, which can form donor and acceptor levels respectively in the electronic structure. Induced geometric defects affect the charge recombination process. Both electronic and geometric defects influence the reactivity of TiO_2 in photocatalytic applications.

478373 Direct Catalytic Coupling and Selective Oxidation of Methane to Ethanol at Low Temperature

Abstract:

The enormous scale of methane reserves has motivated significant research activities focused on its conversion to fuels and chemicals. Since large amounts of natural gas are located in remote areas and transporting gases in pipelines is difficult, processes for converting methane into denser products are highly desirable.

To address this challenge, we developed a bi-functional catalyst that is capable of activating the C-H bond in methane to form surface methyl groups. Some of these methyl groups undergo surface reaction that lead to the formation of higher alkyl chains. In the presence of steam, methanol and ethanol are formed by hydrolysis of the surface alkyl groups. Oxygen has to be co-fed to provide a thermodynamic driving force for this reaction. In addition to alcohols, carbon dioxide and hydrogen are formed as by-products. Importantly, formation of alcohols occurs at 450 °C under steady state conditions with a turnover frequency of at least 50 h^{-1}. This is a significant improvement from previous studies, in which a high-temperature calcination step was required for every turnover.

The identification of promising catalysts and suitable reaction conditions will be discussed. Spectroscopic studies are performed to identify the active sites and the synergistic way in which they enable the reaction.

478711 Studies on Urease Mediated Nanoparticle Formation

Abstract:

Urease is a protein that is found in the Jack Bean, and is commercially purified. It has been reported that urease aids in the process of nanoparticle synthesis[1]. In Dr. Bhagwati Sharma's experiments, urease was found to be catalytic in the synthesis of gold, silver, platinum, and gold-silver alloy particles.[1] Urease is also known to synthesize particles of magnetite, zinc oxide[5], and iron oxide[3]. Dr. Sharma's methods will be tested in these studies in effort to confirm the capability of urease to synthesize silver, gold, and platinum nanoparticles, with the end goal of solving the chemical structure of urease in order to learn more about its properties on a molecular level.

In order to perform the methods that were employed by Dr. Sharma, several different purities of urease will be tested for their effects on the reduction of ionic metals by potassium carbonate. When the correct concentration of urease is used, incubation of each solution will change its color, according to the spectrum of the nanoparticle present: red for gold particles, yellow for silver nanoparticles, and black/brown for platinum nanoparticles. There will be limited or no precipitate present in solution, and the solution will be homogenous. Thus far, $Ag(NO)_3$ has been put through several solutions containing ranges of 2mg/mL Urease to 20mg/mL Urease solution. In some cases, precipitate was formed almost immediately. Since then, the pH has been scaled down to match that of Dr. Sharma's procedure. It is believed that the high alkalinity due to potassium carbonate catalyzed the formation of a precipitate, preventing the solution any opportunity to slowly form nanoparticles.

At the time that the methods have been successfully recreated and the nanoparticles have been formed, the particles will be tested by several methods including cryo-electron microscopy. This will be performed in order to solve for the structure of urease, as well as urease after the particles have been captured. The two structures will be compared in order to evaluate what about the enzyme's structure has facilitated the

nanoparticle capture. This will serve to reveal the mechanism of the reaction.

Nanoparticles have been utilized in the creation of microstructures[1], and are an essential building block for the future of nanotechnology. Cost efficiency of nanoparticle synthesis is extremely crucial for the branching of nanotechnology into other fields of science and technology. The information provided by structural analysis of urease will ideally be beneficial towards the synthesis of more cost efficient enzymes, making particle synthesis less expensive, and therefore making nanoparticles more available to the scientific and industrial communities.

References

1. Sharma, Bhagwati, Sonam Mandani, and Tridib K. Sarma. "Biogenic Growth of Alloys and Core-Shell Nanostructures Using Urease as a Nanoreactor at Ambient Conditions." *Nature.com*. Scientific Reports, 10 Sept. 2013. Web. 2 Oct. 2016.
2. Liang, Zhenpeng, Chaoyang Wang, Zhen Tong, Weihua Ye, and Shiqu Ye. "Bio-catalytic Nanoparticles with Urease Immobilized in Multilayer Assembled through Layer-by-layer Technique." *Science Direct*. Elsevier B.V., Apr. 2005. Web. 2 Oct. 2016.
3. Sahoo, Banalata, Sumanta Kumar Sahu, and Panchanan Pramanik. "A Novel Method for the Immobilization of Urease on Phosphonate Grafted Iron Oxide Nanoparticle." *Science Direct*. Elsevier B.V., May 2011. Web. 2 Oct. 2016.
4. Parashar, Nirala, Upadhyay, Saxena, and Srivastava. "Urease Immobilized Fluorescent Gold Nanoparticles for Urea Sensing." *NCBI*. N.p., May 2015. Web. 2 Oct. 2016.
5. Eghbali M, Farahbakhsh A, Rohani A, Pour A. N. Urea biosensor based on immobilization of urease on ZnO nanoparticles. Orient J Chem 2015;31(2)

478835 DFT Study of the Catalytic Activity of Cu-MFI Zeolite and the Role of the Substitution Sites of Aluminum Atoms

Abstract:

Methane (CH_4) is readily available in gas reserves across the globe. However, its conversion into commodity chemicals, such as methanol or acetic acid is a challenge. Recent works have demonstrated that MFI zeolite modified with Cu^{2+} activates CH_4 at low to moderate temperatures.[1] It is well known that the ionic exchange of Cu^{2+} in zeolites requires aluminum atoms in the framework to compensate the charge difference. In this study, we employ quantum mechanical calculations to analyze how the positions of the aluminum atoms in the framework impacts the catalytic activity of the system.

It has been reported that only three substituted rings have catalytic and adsorption activity in Cu-MFI. These have been identified as a, b, g, and *f* rings, using the nomenclature proposed by Groothaert et al.[2,3] In this work, we systematically analyzed all the possible substitution sites of the aluminum atoms in the a, b, g and *f* rings of the MFI framework. Then the adsorption of methane and several probe molecules, such as NO and CO, on Cu-MFI were analyzed to determine the mechanism of adsorption, the position of adsorbates, as well as to quantify the effect of the position of the Al atoms on the energies of adsorption. The calculations were performed with wB97XD exchange-correlation functional in order to take into consideration the long range interactions of the zeolite. The basis sets used were 6-31+g** for O, Si, Al, and H atoms and SDD for Cu atoms.

Our results demonstrate that the stability of the ringsâ€™ structure might be related to the distance between the Cu^{2+} and the closest Al atom in the framework. It was observed that as that distance increases, the structure is less stable. Ongoing work includes the analysis of adsorption of molecules on the zeolite and the vibrational frequencies of the adsorbates.

478841 Solid Acid Catalytic Dimerization of Ethylene to Butenes

Abstract:

A series of supported $WO_3/ZrO_x(OH)_{4-2x}$ (abbreviated as WO_x/ZrOH) and Ni-$WO_3/ZrO_x(OH)_{4-2x}$ (abbreviated as Ni-WO_x/ZrOH) catalysts was synthesized and examined for ethylene dimerization to isobutene. The catalysts were synthesized by standard aqueous impregnation of ammonium metatungstate onto an amorphous $ZrO_x(OH)_{4-2x}$ support, followed by impregnating $Ni(NO_3)_2$â—$6H_2O$ onto the 10.1-wt%WO_x/ZrOH catalyst. The catalysts were dried and calcined in air at 673-1073K. Characterization with *in situ* Raman spectroscopy revealed the presence of highly dispersed surface WO_x sites and the absence of a NiWOÂ
tested in a fixed-bed reactor and the products were analyzed by online gas chromatography (GC) with a flame ionization detector (FID). The supported tungsten oxide catalysts were found to exhibit low activity for ethylene dimerization. The limited conversion was related to coking of the catalyst surface and the amount of coking closely correlated with increasing nickel loading.

478966 Synthesis and Characterization of Nickel Catalysts for Methane Tri-Reforming

Abstract:

Synthesis and Characterization of Nickel Catalysts for

Methane Tri-Reforming

Nicole Cordonnier

Ohio University

Advisor: Prof. Erdem Sasmaz, Prof. Jochen Lauterbach

The methane tri-reforming process is the combination of steam reforming, dry reforming, and partial oxidation of methane to form syngas with an H_2: CO ratio acceptable for the Fischer-Tropsch process. The tri-reforming process utilizes CO_2 in power plant flue gas, foregoing the usual pre-separation of CO_2. A tri-reforming catalyst must be stable at temperature up to 900°C, have a large surface area capable of storing large amounts of oxygen, and resist formation of coke. Coke formation, as well as sintering and agglomeration cause catalyst deactivation.

In this study, nickel nanoparticles were synthesized for use as catalysts in the tri-reforming process. The catalysts were synthesized using incipient wetness impregnation as well as micro-emulsion techniques to form core shell structures. Catalyst activity and stability were found to depend on both particle size and nonmetal oxide support.

Ni/CeO_2, Ni/MgO, Ni/SiC, and Ni/SiO_2 were synthesized using incipient wetness impregnation. Ni@CeO_2 and

Ni@SiO_2 core shell particles were synthesized using emulsion techniques. Ni@SiO_2 particle sizes were controlled by changing the calcination temperature. Ni/CeO_2 and Ni/MgO were found to have the highest conversions with CO_2 conversions of 54% and 52% respectively, and CH_4 conversions of 98% and 97%. The core-shell catalysts were found to have less activity with a maximum CO_2 conversion of 34% for Ni@SiO_2 with a calcination temperature of 350°C. Lack of core shell activity can be attributed to instability of CeO_2 shell at high reaction temperatures. Particle agglomeration and sintering might also contribute to the deactivation resulting in increased particle size and decreased active surface area.

478700 Functionalized Boron Nanoparticles: Characterization of Thiol-Ene Click Chemistry

Abstract:

Carboranes are a family of cluster compounds comprised of boron, carbon, and hydrogen that are being investigated in a number of applications, including self-assembling monolayers (SAMs), cancer treatment, and non-catalytic synthetic agents. Various functionalities have been attached to carboranes to facilitate their utilization. Thiol-ene â€œclickâ€ chemistry between thiol-functionalized m-carborane and a variety of maleimide derivatives provides a facile path to a wide variety of functionalized boron nanoparticles. The reaction rate shows a strong dependence on the solvent as well as on the position of the thiol group on the carborane. Reaction rates as a function of temperature have been measured with the reaction following second order kinetics. Preliminary testing revealed reactivity differences greater than an order of magnitude between m-carborane-1-thiol and m-carborane-9-thiol.

479020 Design of Highly Efficient Bifunctional Metal-Organic Framework Catalysts for Tandem Catalysis By Shortening the Reaction Pathway

Abstract:

Tandem reactions are important in the scientific community because they reduce time and money to carry out a reaction as well as chemical waste. However, most tandem reactions require two consecutive steps with different catalysts. Metal-organic frameworks have the capability to catalyze reactions; more specifically, the UiO-66 can be functionalized with two different moieties, making them a good candidate for tandem deacetalizationâˆ'Knoevenagel reaction. In this study, it is shown that the UiO-66-type MOF, synthesized using the MHT synthesis method, can be functionalized with both an acidic and basic site at varying amounts. To analyze these affects, a kinetic study using ^{1}H NMR was used to calculate the conversion of benzaldehyde dimethyl acetal (BADA) to benzalmalonitrile.

479056 Selective Dehydrogenation of Ethanol Using Single Atom Alloy Catalysts

Abstract:

This research aimed to improve the process of non-oxidative catalytic ethanol dehydrogenation to produce acetaldehyde and clean hydrogen gas. Acetaldehyde is an important commodity chemical and is widely used as the starting material for the synthesis of a large number of industrial chemicals such as acetic acid, acetate esters, and pentaerythritol. Hydrogen, the other product of this reaction, is also valuable in many industries including clean energy via hydrogen fuel cells. Although the oxidative dehydrogenation of ethanol also produces acetaldehyde, the byproduct is water. Thus, it is attractive to develop new catalysts that are highly active and selective for the non-oxidative dehydrogenation of ethanol.

Small amounts of platinum, palladium, and nickel were added to the surface of Cu/SiO_2 nanoparticles to enhance the dehydrogenation activity of copper. Pd and Pt atoms are singly dispersed in Cu, forming single atom alloys (SAAs), whereas Ni atoms are highly dispersed in Cu, forming a highly dilute alloy. Our catalytic measurements show that forming PdCu SAAs does not improve the catalytic performance of copper. PtCu SAAs improve the stability of the copper catalyst, but have little effect on the activity. Interestingly, creating highly dilute alloys of Ni in Cu significantly improves both the activity and stability of the copper catalyst. Further study and characterization of the NiCu highly dilute alloy is underway, as this will help us understand the reaction mechanism and identify the active phase, thus guiding the design of new catalysts that further improve the dehydrogenation reactivity.

479084 Surface Spectroscopic Study of Renewable Ethylene Glycol Synthesis over Silicon Dioxide-Supported Metal Catalysts

Abstract:

Ethylene glycol (EG) is an extremely versatile chemical and, thus, is constantly in high demand. The current method of producing ethylene glycol involves oxidizing ethylene from natural gas into ethylene oxide over a silver (I) oxide catalyst, followed by hydration using acid. At nearly $37 per gram, however, silver (I) oxide is not very affordable. Recently, a new and economical synthetic method has been proposed using silicon dioxide-supported metal catalysts and syn gas (a mixture of carbon monoxide and hydrogen gas) .In a coupling reaction between carbon monoxide and methyl nitrite catalyzed by the supported metal, dimethyl oxalate (DMO) and nitric oxide are formed. The hydrogen in the syn gas can then be used to catalytically reduce DMO into ethylene glycol and methanol. Finally, the nitric oxide and methanol can be oxidized with atmospheric oxygen to regenerate methyl nitrite, which can then be recycled back into the reaction.

This experiment tested the selectivity and stability of SiO_2-supported copper during the reduction of DMO into ethylene glycol. Catalysts were prepared in 15-20 percent weight copper, using incipient wetness impregnation, as well as urea-assisted gelation. Fourier transform infrared spectroscopy was then performed as DMO dissolved in ethanol flowed over the catalyst. These spectra allowed some of the functional groups forming on the surface of the metal catalyst to be identified, based on shapes of peaks at certain wavenumbers. Determining the identities and concentrations of byproducts will give insight into the selectivity of the Cu/SiO_2 catalyst. Taking many spectra over extended periods of time will show how stable the catalyst is under reductive conditions.

479236 Catalytic Materials for Demetallation of Algal Biocrude

Abstract:

As our planetâ€™s energy resources grow scarce, we must become cultivators of energy. Algae is a renewable resource that has fast growth rate and characteristics that point to it being favorable for the production of liquid fuel. However, algae, and its produced crude bio-oil, stores high concentrations of metals, especially iron, in its composition, which produces problems in petroleum refineries, as they are not designed to process iron infused feedstocks. Therefore, different types of catalysts, such as CoMo/silica-alumina, activated carbon, Ru/C and Ni/silica-alumina, were used under different experimental conditions to obtain the one with best demetallation properties and biocrude yield. To obtain our biocrude, batch-mini reactors are loaded with water, algae, catalyst, and enter a pre-heated sand-bath for hydrothermal liquefaction (HTL). After cooling down, the reactorsâ€™ contents are extracted with a solvent and then centrifuged to obtain different phases of products: organic, solid and aqueous. The organic phase, where the biocrude exists, is extracted and dried. Once the biocrude yield is known, the biocrude is dissolved with another solvent and run through the inductively coupled plasma optical emission spectrometry machine to know the metal composition contained in the biocrude. The solid and aqueous phases are also analyzed, using multiple characterization techniques, in order to obtain a profile of where all the metals are being deposited after HTL. It was found that the best experimental conditions in order to obtain the best demetallation and biocrude yield were using: methyl tert-butyl ether as a solvent, temperature of 400ËšC, an hour of HTL, and more amount of catalyst. It was also found that, so far, the best catalyst for algal biocrude demetallation is Ni/silica-alumina, and that most metals after HTL are deposited in the solid phase. However, many more catalysts and characterization techniques must be employed in the future in order to gain more information and finite results for this area of research.

479334 Catalytic Application of Few Layer MoS_2

Abstract:

Catalytic Application of Few-Layer MoS_2

***Ryan Dekle, **Swagotom Sarkar, **Hongmei Luo,* Sanchari Chowdhury**

***New Mexico Tech, **New Mexico State University**

The reduction of 4-Nitrophenol by sodium borohydride is studied as a model reaction to test the catalytic activity of defect free and defect rich annealed and non-annealed few-layer MoS_2. With increasing demand for 4-nitrophenol and its derivatives, as raw materials for agrochemicals, dyes, drugs, polymers, rubbers and different natural products, the reduction and removal of 4-nitrophenol has drawn extensive attention in recent years. This reduction is often considered as a benchmark reaction to test a catalyst nanoparticle and the progress of the reaction can be conveniently monitored by UV-visible spectrophotometry. We observed at room temperature, there is negligible reduction with $NaBH_4$ only. However in the presence of MoS_2 the reaction is quite efficient. Defect free few layer MoS_2 have shown efficient catalytic activity towards 4-

Nitrophenol reduction. Since hydrothermally synthesized few layer MoS_2 does not show 1T phase, the catalytic activity can be attributed to its metallic edge. Defects such as vacancies are usually expected to be active centers for molecular adsorption and chemical reaction and often more catalytic sites can be activated by introducing sulphur (S) vacancies. However in our case the defect rich samples show lower catalytic activity than that of defect free samples. This may be due to hindered electron transport in MoS_2 due to presence of defect.

479550 Site and Structural Changes to Sn-Beta Zeolites in Aqueous Media and Their Consequences for Glucose Isomerization Catalysis

Abstract:

As the pressure to reduce greenhouse gas emissions increases, biofuels provide a promising alternative to fossil-derived fuels due to their overall lower net CO_2 production. One biofuel production pathway involves the conversion of sugars from cellulosic biomass sources, and the isomerization of glucose to fructose is an important reaction step in this pathway. This reaction can be catalyzed by isolated tin atoms within the framework of Beta zeolites (Sn-Beta) [1], in which the proposed active site is a hydrolyzed Sn site in open coordination within the framework ($(SiO)_3$-Sn-(OH)) [2, 3]. Here, we report the effects of exposing Sn-Beta catalysts to aqueous media at high temperatures (373 K), prior to measuring glucose isomerization rates. Catalysts exposed to water at 373 K for 0-60 minutes prior to reaction showed rates (per total Sn) that were up to 2.6 times higher than on catalysts that were not exposed to water prior to reaction. Longer exposure times to water (>60 minutes) prior to reaction resulted in rates (per total Sn) that decreased with increasing exposure time, indicating catalyst deactivation. In order to determine how the catalyst structure changed during these activation and deactivation periods, pyridine and deuterated acetonitrile infrared spectroscopy were used to quantify the types of Sn sites present on the catalyst before and after hot water treatment. Among the different samples studied, the magnitude of the rate enhancement (per total Sn) observed after hot water treatment increased systematically as the fraction of open Sn sites (per total Sn) present initially on the catalyst decreased. This suggests that the activation period forms new hydrolyzed open Sn sites in the presence of water that become available as active sites for the reaction. Hydrolysis of framework bonds caused long-term catalyst deactivation, however, resulting from both tin leaching and defect silanol group formation. These findings are consistent with open Sn sites behaving as the dominant active site for glucose isomerization in Sn-Beta, and provide new insights into the effects of water exposure on site and structural changes to Sn-Beta zeolites.

[1] M. Moliner, Y. RomÃ¡n-Leshkov, and M.E. Davis, *Proc. Natl. Acad. Sci. USA* 107 (2010) 6164â€"6168.

[2] R. Bermejo-Deval, *et al.*, *ACS Catal.* 4 (2014) 2288-2297.

[3] J.W. Harris, *et al.*, *J. Catal.* 335 (2016) 141-154.

479626 Investigation of Fluidized Bed Agglomeration of Cu-Based Oxygen Carriers

Abstract:

Combustion of fossil fuels creates environmental concerns due to the emission of flue gases causing air pollution and contributing to global warming. Chemical Looping Combustion (CLC) is a proposed alternative technology for production of energy that has advantages over conventional combustion of fossil fuels, particularly with regard to emission of CO_2. The CLC process employs two interconnected fluidized bed reactors. Air fed to the first reactor (air reactor) oxidizes metal-based "oxygen carrier" particles, which then circulate ("loop") to a second reactor (fuel reactor) where oxygen (O_2) is released from the oxygen carrier and combusts the fuel injected in the fuel reactor. The process enables isolation of CO_2 since the fuels are combusted in an O_2-containing environment in the absence of nitrogen. Chemical looping combustion with oxygen uncoupling (CLOU) is a derivative of CLC for combustion of solid fuels through specific metal based oxygen carriers with high capacities to release O_2.

Different metal based carriers have been developed for CLOU, and copper based oxygen carriers have shown promising results for oxygen liberation and particle strength in the fluidized bed reactors compared to other metal based particles. In the air reactor, cuprous oxide (Cu_2O) reacts with O_2 in air to form copper oxide ($Cu_2O + \frac{1}{2} O_2 \leftrightarrow 2\ CuO$). The reverse reaction happens in the fuel reactor where O_2 is released and cuprous oxide is regenerated and recycled again to the air reactor. The vigorous fluidized bed environment and high temperatures required for efficient fuel combustion poses risks of agglomeration of the particles, where particles adhere to each other to form large lumps causing collapse of the fluidized bed and subsequently costly shutdown of the system.

Agglomeration testing in a fluidizedbed system has been conducted to find suitable operating windows in

terms of temperature and superficial velocity for efficient conversion in the air and fuel reactors. Oxygen capacity and absorption and release rates of a set of copper based particles have been evaluated while monitoring fluidization behavior of the particles for potential detection of agglomeration. A differential pressure measurement across the reactor has been found to be valuable for detecting agglomeration when particles start to defluidize in the bed. Higher superficial velocities, lower bed loadings and lower copper loading on the particles were found to reduce risks of agglomeration. This presentation emphasizes different techniques employed for agglomeration testing and detection. Additionally, optimum operating conditions for the carrier particles in fluidized beds will be presented in terms of temperature and superficial velocity.

479738 Heterogeneous Catalyzed Esterification of Caproic Acid and Lauric Acid with Isoamyl Alcohol from Fusel Oil

Abstract:

The industry of bioethanol produces a by-product known as fusel oil, which represents around 0,5 vol% of the production of bioethanol and which contains mainly amyl alcohol isomers [1]. These alcohols represent 41-85 vol% of fusel, depending on the raw materials and operating conditions of the fermentation step [2]. One of those isomers, isoamyl alcohol, is a raw material in the production of value added esters largely used as flavors, fragrances, plasticizers and bio solvents in food, cosmetic and pharmaceutical industry. There are some esters that also exist in the heaviest fraction of fusel oil, such as isoamyl caproate and isoamyl laurate; isoamyl caproate is used in green apple and pineapple-like flavor formulations and on the other hand, isoamyl laurate smells like peach and wine.

The aim of this project is to carry out a kinetics study of the production of isoamyl caproate and isoamyl laurate using Amberlyst 35WET cationic exchange resin as catalyst and stirring speeds of 600 rpm and 700 rpm. The experimental design used is Box-Behnken for three factors: temperature, reactants molar ratio and catalyst loading. All the samples from the reacting system were analyzed by acid value determination. Reaction temperature was varied between 90°C and 110°C, and the catalyst loading was studied in the range between 0,5% and 3% m/m of the reacting mixture.

In the two cases one of the reactants is put in excess. For isoamyl laurate reaction isoamyl alcohol is the excess reactant and, on the contrary, for isoamyl caproate reaction this is used as limiting reactant in order to prevent two liquid phases in the reactor. Molar ratios of 2:1, 3:1 and 4:1 are planned in the experiment design, however until now only the molar ratio 2:1 has been analyzed.

In both cases, the reactions achieve conversions higher than 90% for temperatures above 100°C but reaction time is larger for isoamyl caproate production (8-9 hours) than isoamyl laurate production (6-7 hours). Reactions are strongly temperature dependent rather than the catalyst loading. Changes of temperature from 90°C to 100°C represent changes of 40% in conversion while changes of catalyst loading between 0,5% to 3% represent changes only of 4% in conversion. It can be concluded that it is not necessary to use large amounts of resin to improve the process yield.

References

[1] Z. KÜCÜK y K. CEYLAN, «Potential Utilization of Fusel Oil: A Kinetic Approach for Production of Fusel Oil Esters Through Chemical Reaction,» *Turkish Journal of Chemistry,* nº 22, pp. 289-300, 1998.

[2] F. Leyva, Á. Orjuela, A. Kolah, C. Lira, D. Miller y G. Rodríguez, «Isoamyl propionate production by reactive distillation,» *Separation and Purification Technology,* nº 146, pp. 199-212, 2015.

479961 Solvent Effects during Acid-Catalyzed Esterification Reactions

Abstract:

The combined use of chemical and biological catalysis will potentially allow for the sustainable production of high value chemicals that are traditionally sourced from petroleum. Acid catalyzed esterification is one reaction that is important for the production of these higher value chemicals. This study explored the solvation effects of a polar solvent (THF) versus those of a nonpolar solvent (toluene) on esterification reactions catalyzed with Amberlyst-15 and beta zeolite.

The reaction was fractional order with respect to butanol concentration, and first-order with respect to butyric acid concentration, when using toluene as a solvent. In contrast, the reaction in tetrahydrofuran was first order with respect to both butanol and butyric acid. This discrepancy can be explained based on solvation of abundant surface intermediates leading to the appearance of a site blocking term in the rate expression for toluene. Accompanying the change in the reaction order, the reaction rate was increased by approximately an order of magnitude when using toluene as a solvent. These observations can be explained in the context of transition state theory applied to thermodynamically nonideal systems.

479978 Characterization of ALD Fuel Cell Catalysts Using Durability Testing and Electrochemical Impedance Spectroscopy

Abstract:

One of the barriers to commercialization for proton exchange membrane fuel cells is loss of activity of the cathode catalyst over time. This is caused by degradation in the form of particle migration and sintering as the fuel cell is subjected to harsh voltage cycling. Atomic layer deposition (ALD) utilizes a self-limiting reaction to produce uniform coatings on a particle substrate. Using ALD, a TiO2 layer was deposited on a commercial catalyst to investigate its effectiveness in preserving catalytic activity. Preliminary results from durability tests indicate improved stabilization of the catalyst in the form of a lower rate of loss of electrochemical surface area (ECSA). To further characterize the performance of the coated catalyst, the impact of the TiO2 layer on specific resistancesâ€"namely mass-transfer and charge-transfer resistancesâ€"at the cathode is being investigated using electrochemical impedance spectroscopy (EIS). In EIS, the catalyst is subjected to small sinusoidal perturbations around select voltages. The current response is measured at each voltage to find the impedance behavior, which is presented in a Nyquist plot. Equivalent circuit diagrams are constructed and fitted to extract quantitative information about resistances in the cathode half-reaction. From this information, the TiO2 layer can be adjusted to further improve performance at the cathode.

479784 Exchange Kinetics of Quinolinyl Ketones with Boronic Acids Via Rhodium Catalyzed C-C Bond Activation

Abstract:

Carbon-carbon Ïƒ bond activation is difficult due to the strength of the Ïƒ bond and the steric hindrance around the bond. Using a quinolinyl ketone system, however, activation can be achieved with a rhodium catalyst, allowing not only carbon-carbon Ïƒ bond activation, but also functionalization using boronic acids. The exchange of aryl substituents between quinolinyl ketones and boronic acids, proceeding through carbon-carbon Ïƒ bond activation, is the subject of a kinetic study. NMR and kinetic studies are used to determine the rate law of the reaction using ortho-fluoro quinolinyl ketone and 4-trifluoromethyl boronic acid. NMR studies are used, in addition, to compare the rates of various aryl substituents of both boronic acids and quinolinyl ketones to ortho-fluoro quinolinyl ketone and 4-trifluoromethyl boronic acid exchange.

480441 Redox Active Material Development and Testing for Efficient Solarthermal Water Splitting

Abstract:

Solar thermal water-splitting (STWS) is a potential solution for global climate change and energy dependence issues as it is a renewable and carbon neutral source of hydrogen, an energy-dense fuel. While active materials already exist for STWS, continued material development for improved hydrogen production capacity, faster kinetics, and long-term stability is needed to improve the feasibility of STWS on a large scale. Perovskites (ABO_3) and spinels (AB
have shown promise as active materials for improving the efficiency of solarthermal water-splitting. The complex perovskite $Sr_{0.4}La_{0.6}Mn_{0.6}Al_{0.4}O_3$ (SLMA), the hercynite ($FeAl_2O_4$) spinel, and the most well-studied STWS material ceria have all been separately shown to produce hydrogen, but comparisons between these studies are difficult due to different experimental conditions. These three materials were tested under the same conditions and compared for both thermodynamic and kinetic viability. Novel binary perovskites and spinels were also chosen using an efficient screening method with density functional theory. These materials were synthesized, fully characterized, and tested for hydrogen production capacity and redox kinetics using a stagnation flow reactor and thermogravimetric analysis.

480463 Characterization and Stability of Metal Oxide Supported Cobalt Nanorod Catalysts for CO_2 Hydrogenation

Abstract:

Converting carbon dioxide to useful chemicals, such as hydrocarbons, through hydrogenation processes has attracted great deal of interests recently amid the global rising of CO_2 levels. Cobalt nanorods are recently found by our group to have high activity to CO_2 hydrogenation and high selectivity towards hydrocarbons due to their preferentially exposed {110} facet on the surface. To translate these cobalt nanorods into a practical catalyst for industrial plants, we developed a method to preserve the surface faceting when supporting the nanorods with a metal oxide. Charged ligands were grafted onto the nanorods to enable the metal oxides to grow around the nanorods. Three different metal oxides â€" silica, alumina, and titania â€" were grafted as different supports onto cobalt nanorods. The mesoporous structure of the supports also allows for improved heat and mass transfer during hydrogenation. Using Induced Coupled Plasma (ICP), the synthesized catalysts were determined to be equal parts metal oxide support and cobalt oxide catalyst. The alumina and titania supported cobalt nanorods indicated improved activity toward hydrogenation over pure cobalt nanorods, and the silica supported nanorods exhibited the same catalytic performance. The enhancement of catalytic activity was further investigated with Temperature Programmed Reduction (TPR) experiments, which indicate that a strong interaction between cobalt nanorods and oxide support is the main cause of activity increase. The reactivity and selectivity toward hydrocarbon formation were not hindered by supporting the cobalt nanorods. Furthermore, the structures of the pure and supported catalysts were characterized before and after the reaction to determine stability and preservation of the {110} facet. From X-Ray Diffraction (XRD) analysis and Raman spectroscopy, the supported catalysts were determined to have the cobalt oxide spinel structure before and after the reaction. The Raman spectra also revealed there was coke formation during the reaction. From nitrogen adsorption/desorption data, the coking resulted in the surface area and the pore size of the unsupported and supported catalysts decreasing.

480499 Synergistic Interactions Between Copper and Iron Oxide during the HT-WGS Reaction

Abstract:

The supported Fe-Cu-/SiO_2 catalyst system investigated as a model catalyst for examining the synergistic interaction between copper and iron during the High Temperature Water-Gas Shift (HT-WGS) reaction. The catalysts studied were: bulk Fe_2O_3, supported 5%Fe/SiO_2, supported 5%Cu/SiO_2, supported 5%Cu5%Fe/SiO_2, supported 10%Cu5%Fe/SiO_2, supported 10%Cu/SiO_2, supported 20%Cu/SiO_2, and supported 20%Cu5%Fe/SiO_2. The catalysts were studied for the steady-state HT-WGS reaction to determine catalyst activity as a function of catalyst composition and reaction temperature. There was a measurable increased activity for catalysts with containing both copper and iron revealing a synergistic interaction between these components. The reducibility of the catalysts was reducibility was probed with CO-Temperature Programmed Reduction (TPR) that revealed that synergistic interaction between Cu and Fe was related to copper aiding the reduction of the Fe_2O_3 phase.

480523 Partial Oxidation of Methane on Nickel Carbide Catalysts

Abstract:

Partial oxidation of methane is a process used industrially to convert methane to syngas, a mixture of hydrogen and carbon monoxide. Syngas is then used in chemical syntheses such as hydrogen and methanol, or upgraded through the Fischer-Tropsch process to produce longer-chained hydrocarbons, and ultimately products like gasoline. Rhodium is commonly used as a catalyst for this process; however, rhodium is prohibitively expensive. Due to this expense, a new catalyst for partial methane oxidation is desirable. Nickel has been investigated as this alternative catalyst, and is in fact used in some industrial processes. However, lower activity, deactivation by oxidation, and formation of coke are problems associated with the use of metallic nickel as the catalyst. In separate studies, metal carbides have been shown to have catalytic properties similar to that of noble metals, a class into which rhodium falls.

In this work, the effectiveness of nickel carbide as a catalyst for partial methane oxidation is investigated. Nickel carbide nanoparticles were synthesized and characterized by XRD and TEM, and found to be nickel carbide particles with diameter of approximately 30 nm. These nanoparticles were then used in a temperature programmed reduction (TPR), and reacted under atmospheres of varying Ar, CH_4, and O_2 composition under varying temperature ramps ranging from nominal temperatures of 20°C to 800°C. After pretreatment under Ar at 350°C, the nickel carbide showed evolution of hydrogen starting near 460°C. Under a mixture of Ar, CH_4, and O_2, oxygen consumption and hydrogen production started around 300°C. The characterization of the nickel carbide after the reaction showed agglomeration of the nickel carbide particles and degradation of the nickel carbide phase and associated formation of a metallic nickel phase. Further results of the methane partial oxidation activity and stability of the nickel carbide catalyst will also be presented and discussed.

480624 Electrochemical Reduction of Carbon Dioxide to Formic Acid in Ionic Liquid [Emim][N(CN)2]/Water System

Abstract:

The electrochemical reduction of CO_2 to formic acid or formate is a promising technology for addressing transfer and storage issues of current intermittent renewable energy resources. One of its main challenges is that the faradaic efficiency is low in the inorganic salt aqueous system because of the hydrogen evolution reaction is much easier than CO_2 electroreduction on most metal electrodes. Previous investigations have revealed that an ionic liquid system can provide desired products with the high faradaic efficiency and eliminate the hydrogen evolution reaction. However, this previously used ionic liquid system has a major pitfall because there is a net consumption of the ionic liquid in its reaction mechanism. In this manuscript, we present an alternate route, where 1-ethyl-3-methylimidazolium dicyanamide ([Emim][$N(CN)_2$]) aqueous solution is used to increase the solubility of CO_2 and suppress the water electrolysis without being consumed in the reaction. Cyclic voltammetry is performed in [Emim][$N(CN)_2$] aqueous solution on the Sn power electrode to determine the on-set potential of CO_2 electrolysis and its electrochemical stability. In addition, CO_2 electrolysis is carried out at various fixed potentials and concentrations of electrolyte to understand their effects on the faradaic efficiency. The maximum faradaic efficiency obtained during electrolysis is 81.9% towards the formation of formate at 1.8 V vs. Ag/AgCl and 0.5 mol L^{-1} of ([Emim][$N(CN)_2$].

480640 Size Control of Colloidal Pd Nanoparticles: Role of Ligand Binding Strength

Abstract:

Through a rapidly developing interest in nanotechnology, the application of Palladium nanoparticles has been realized in several areas. The catalytic properties of Pd nanoparticles are one of their most researched applications. They are commonly used in the early stages of pharmaceutical drug manufacturing, specifically for drugs that are synthesized through organic cross-coupling reactions. To synthesize particles of specific size, the particles must often be stabilized by ligands. Studies have demonstrated that nanoparticle size is directly controlled by the type and concentration of ligand, but it is still unclear if they prevent nanoparticle agglomeration or have a role in controlling growth via blocking of surface sites. Our hypothesis is that nucleation and growth kinetics are affected by the ligand-metal binding strength, which can be controlled by the type of ligand and solvent. In this investigation, we aim to determine the relationship between Pd nanoparticle size and the binding strength of phosphine and amine ligands to the Pd precursor (PdClâ,,). Synthesis reactions were carried out with $PdCl_2$ in pyridine and $NaBH_4$ as the reducing agent, in addition to the various capping agents. Particles sizes were measured by small angle X-ray scattering (SAXS) and transmission electron microscopy (TEM) and ranged from 1 nm â€" 4 nm in diameter. The binding energy between the precursor and the ligand was measured using isothermal titration calorimetry (ITC). Using ITC, we are able determine the stoichiometry, binding heat, and equilibrium constant for the binding between different ligands and the precursor, and in turn better understand their binding behavior. The results are used to correlate the thermodynamics of metal-ligand interaction with the final size of the nanoparticles to help predict the size using other ligands and ligand:Pd ratios.

480662 Synthesis of Platinum/Ceria Catalyst Using Supercritical Fluid Deposition

Abstract:

Catalysts containing platinum and ceria on alumina support are very important in environmental and energy applications, as they are very efficient at converting CO, hydrocarbons, and nitrogen oxides (NO_x) to less harmful CO_2, H_2O, and N_2. They are useful for the water-gas shift reaction, the goal of which is to oxidize CO in the presence of water to form CO_2 and H_2, which can be used in fuel cells for power generation. Dispersion of the active metal sites is key in evaluating catalyst activity, as it measures the efficiency of use of metal atoms. Dispersion is defined as the fraction of the number of metal atoms exposed over the total number of metal atoms. Highly dispersed catalysts are the most effective for the water-gas shift reaction. A promising synthesis for these catalysts is supercritical deposition. This technique leaves metal precursors on the surface, which are reduced to highly dispersed Pt in post-adsorption treatments, leaving homogeneous distributions of surface nanoparticles. We have found that exposure to CO_2 causes supported ceria to become highly dispersed, but, after calcination, the ceria accumulates around platinum. The effect of exposure to supercritical CO_2 on the nanostructure of deposited materials was studied. Catalysts were synthesized using incipient wetness impregnation, supercritical deposition, and incipient wetness followed by exposure of adsorbed precursors to supercritical CO_2. Dispersion measurements using CO pulse chemisorption at 195 K indicate that supercritical fluid deposition produced the most highly dispersed catalysts, followed by incipient wetness impregnation, with subsequent CO_2 exposure yielding the lowest dispersions.

480700 Novel Synthesis of CHA Zeolites

Abstract:

Zeolites are used the world over for applications that span all walks of industry. Initially popular for their ion exchange properties, new artificial processes allow for synthesis of high quality crystals; zeolites can now be used for precise separation and high conversion catalysis. Some zeolites, while occupying large portions of usage markets are expanding their portfolio of uses. These newer uses for zeolites require research and highly reliable methods of production for development. The limitation on the development and usage of the more advantageous zeolites is the excessive cost of manufacturing required to produce even small amounts of pure crystals. Many methods for synthesis exist, some that work extremely well may have specific limitations that exclude them for use in larger industrial applications. It is hydrothermal crystallization that is the most commonly used for the more prominent zeolite frameworks, as the process allows for greater control of product characteristics. This work examines the kinetics and reproducibility of synthesis of CHA using a new structural directing agent and several different sources of silica and alumina. This new method is significantly cheaper than the methods widely accepted for manufacture of CHA type and can produce zeolites with custom Silica: Alumina ratios.

480268 Catalytic Conversion of Bioethanol to 1,3-Butadiene

Abstract:

Determining the fundamental details of the catalytic conversion of bioethanol to butadiene, a poorly understood reaction at present, is needed in order to guide the rational design of advanced green catalysts for the bioethanol conversion to butadiene for the manufacture of green tires. Silica-supported ZrO_2 and ZnO catalysts were synthesized to study the mechanism and kinetics behind the conversion of bioethanol to 1,3-butadiene. The ZrO_2 catalysts were synthesized by incipient wetness impregnation from toluene solutions of zirconium (IV) n-propoxide onto a SiO_2 support. The ZnO catalysts were prepared by incipient wetness impregnation from aqueous solutions of zinc acetate dihydrate onto a SiO_2 support. The catalysts were dried overnight and calcined in air at 500°C. The catalytic activity of each catalyst was tested in a fixed-bed reactor and the products were analyzed by gas chromatography (GC) with a flame ionization detector (FID). Several experiments were conducted to study the kinetics which include time to achieve steady state, and temperature/pressure effect on ethanol conversion. The surface kinetics of the ethanol conversion pathways to 1,3-butadiene were also investigated with Temperature Programmed Surface Reaction (TPSR) spectroscopy.

The findings will be presented and discussed.

480724 Investigating Ligand-Nanoparticle Interactions and Catalytic Activity through Organic-Aqueous Tunable Solvents Recovery

Abstract:

Dispersed gold nanoparticles have greater catalytic activity than those deposited on a fixed support. Current methods to recover dispersed nanoparticles require either specialized synthetic methods or produce large quantities of solvent waste and energy. These methods are difficult to scale up and can affect the morphology of the nanoparticles resulting in decreased catalytic performance. We propose a recovery system that utilizes Organic-Aqueous Tunable Solvents (OATS): a homogeneous mixture of water and an organic solvent selected specifically for its ability to absorb CO_2. The mixture undergoes a phase separation under pressure from CO_2 to form a biphasic system consisting of water and a CO_2-expanded organic solvent. The ligand selected to disperse the gold nanoparticles was Poly(vinylpyrrolidone) (PVP) because it is water soluble and binds weakly allowing surface access to the reacting species. Gold nanoparticles of size 9.4 ± 1.4 nm were synthesized and subjected to four thermal treatments (no thermal treatment, 40, 50, and 60°C) as well as pressure treatment between 1-11.3 bar of absolute CO_2 (the pressure range at which phase separation in OATS occurred). Complete recovery of gold nanoparticles was achieved as determined by UV-vis extinction spectra of the localized surface plasmon resonance for each of the four treatments. The hydrogenation reaction of 4-nitrophenol was used to determine the catalytic activity of the nanoparticles. Catalytic activity decreased after the nanoparticles underwent thermal treatment or pressure recovery but remained relatively high. Reduction of catalytic activity can be linked to loss of active sites which is likely due to the passivation of the nanoparticle surface resulting from increased PVP carbonyls reaching the surface. Surface bound ligands prevented precipitation and growth in OATS, but they also slowed dynamic surface restructuring which decreased catalytic activity.

480727 Engineering Bilayer Electrolytes in Solid Oxide Electrolysis Cells

Abstract:

Solid Oxide Electrolysis Cells (SOECs) are solid-state electrochemical devices that under an applied potential can convert low-energy molecules (e.g., CO_2 and H_2O) into high-energy molecules (e.g., CO and H_2, respectively) and gas phase oxygen. In SOECs, the cathode compartment is responsible for the reduction reaction, which in the case of CO_2 electrolysis, leads to formation of CO and oxygen ions. Oxygen ions then diffuse through an oxygen-conducting solid oxide electrolyte, (traditionally yttria stabilized zirconia (YSZ)) to the SOEC anode, where oxygen evolution occurs forming gas-phase oxygen. The standard SOEC cathode is usually composed of a metal electrocatalyst (e.g., Ni) over a ceramic oxide (e.g., YSZ), while the anode is composed of strontium-doped lanthanum manganite (LSM) over the same ceramic oxide. YSZ is traditionally used as the solid oxide electrolyte, yielding viable and durable SOECs, but limits the device operation to temperatures above 800°C due to its low oxygen conductivity. One way to overcome this challenge is to replace YSZ with gadolinium doped ceria (GDC), which exhibits higher ionic conductivity, thus promoting lower operating temperatures. However, in a reducing atmosphere GDC shows mixed ionic and electronic conductivity, yielding open circuit potential values that are generally lower than the theoretically calculated voltage, thus reducing the cell performance. We propose a method for eliminating electron leakage in GDC by applying a very thin layer of YSZ over the GDC electrolyte that acts as an electron insulator, allowing only the oxide ions to pass through the electrolyte. We show that the coated GDC SOEC yields improved electrochemical performance over both the YSZ and GDC only electrolyte cells.

480728 Analysis of the Electrocatalytic Activity of Nickelate Oxides As Cathode Electrocatalysts for Li-Air Batteries

Abstract:

Analysis of the Electrocatalytic Activity of Nickelate Oxides as Cathode Electrocatalysts for Li-Air batteries

Montserrat Diaz, Ayad Nacy, Samji Samira and Eranda Nikolla

Department of Chemical Engineering and Materials Science, Wayne State University, Detroit, MI, 48202

Li-air batteries are among the most promising energy storage technologies due to their high theoretical energy density (11,680 Wh/kg), rivaling that of gasoline (13,000 Wh/kg).[1, 2] One of the main challenges with Li-air batteries include the high overpotential losses due to the electrochemical reactions occurring on the cathode side [$2Li + O_2 \leftrightarrow Li_2O_2$], where precise reaction pathways remain undefined. It has been shown that the kinetics of oxygen reduction and oxygen evolution reactions (ORR and OER) can be improved by incorporating active catalysts on the cathode surface.[3] Among these catalysts, noble metal nanoparticles (such as gold[4], platinum[4] and palladium[5]) supported on porous carbon cathodes have shown superior performance. However, these metals are expensive and scarce. Therefore, the development of cost effective cathode electrocatalysts for Li-air batteries remains a challenge.

In this contribution, we show our efforts in designing robust non-precious metal based heterogeneous catalysts for Li-air cathodes. We find that the incorporation of layered nickelate oxide catalysts in Li-air cathodes significantly increases the decomposition rate of Li_2O_2 during the charging process due to their ability to lower the activation barrier for the oxygen evolution reaction.[6] We have also analyzed the effect of the chemical composition of nickelate oxides on the activity toward the electrochemical dissociation of Li_2O_2 with the aim of tuning their performance. The composition effects on the activity of nickelate oxides will be discussed.

[1] Imanishi, N; Yamamoto O. *Material Today* **2014**, 17, 1.

[2] Luntz, A. C.; McCloskey, B. D. *Chem Rev* **2014**, *114*, 11721.

[3] Girishkumar, G.; McCloskley, B.; Luntz, A.C; Sawson, S.; Wilcke, W *J. Phys. Chem. Lett.* **2010**, 1, 2193-2203

[4] Lu, Y. C.; Xu, Z. C.; Gasteiger, H. A.; Chen, S.; Hamad-Schifferli, K.; Shao-Horn, Y. *J Am Chem Soc* **2010**, *132*, 12170.

[5] Lei, Y.; Lu, J.; Luo, X. Y.; Wu, T. P.; Du, P.; Zhang, X. Y.; Ren, Y.; Wen, J. G.; Miller, D. J.; Miller, J. T.; Sun, Y. K.; Elam, J. W.; Amine, K. *Nano Lett* **2013**, *13*, 4182.

[6] Nacy, A.; Ma, X. F.; Nikolla, E. *Top Catal* **2015**, *58*, 513.

480753 Exploration of Viable Methods to Form 1-Thia-3,4-Diazolidine-2,5-Dione (TDAD) Based Polymers

Abstract:

Novel, well-designed photodegradable polymers have the potential to profoundly influence society due to their wide range of applications. The photochemistry and bisfunctionality of the 1-thia-3,4-diazolidine-2,5-dione (TDAD) structure allows for the development of these unique materials. An area of particular interest with TDAD polymers is that of photoresists. Photoresists are light sensitive polymers used to coat silicon wafers that are etched to create integrated circuits, which are a vital component of modern electronics. Portions of the photoresists are photolyzed, and then a wash is required to remove the remaining organic molecules. Etching the exposed sections of the silicon wafer creates the semiconductor of the integrated circuit. If a photodegradable polymer used in this process photolyzed completely to innocuous gaseous molecules, the wash step to remove the products of the photolytic decomposition could be eliminated. The cost, waste, and time associated with this wash would therefore be unnecessary, which would have a massive impact upon the industry.

Initial experiments with TDADH utilizing base-catalyzed anion polymerization have shown that with efficient leaving groups, the intermediate TDAD anion can rearrange, leading to a thermal decomposition. Experiments using 1,2-diiodoethane have exhibited this decomposition, which is partially due to the stability of the ethylene decomposition product. New methods of forming these polymers which will avoid this rearrangement are being examined.

For instance, the method of using 1,1-disubstituated dihalides instead of 1,2-disubstituted dihalides in the anionic polymerization is being explored. Decomposition of the intermediate TDAD anion made from this type of dihalide would afford high-energy carbenes, making this undesirable thermal rearrangement unlikely.
In the latest polymerization effort diiodomethane was used in the alkylation of the TDAD salt, as it is a 1,1-dihalide that would minimize the likelihood of the undesirable rearrangement. Solubility tests indicated that solids produced by this reaction were soluble in dimethyl sulfoxide but insoluble in chloroform. Importantly, some of the solid produced is insoluble in water. The combination of insolubility in less polar organic solvents, coupled with solubility in the very polar DMSO and insolubility in water is positively indicative of the presence of a TDAD polymer or oligomer. The TDAD polymer with just a CH_2group between the rings may be so polar that it is water soluble which would explain the partial solubility the products show.

To eliminate confusion 1,1-dibromoheptane has been synthesized, which will be used to create a TDAD polymer with a long organic chain. This nonpolar group will decrease water solubility of the resulting TDAD polymer. The solubility tests would be made clearer and the carbon chain would give the polymer more groups that could be examined spectroscopically. Because TDAD polymers offer the possibility of different connecting groups a variety of syntheses are possible. Besides experimenting with alternative leaving groups and manipulating the alkylation processes as mentioned, nylon 6 approach, acid-catalysis and ring opening metathesis-type polymerizations will be explored to work towards the goal of developing a photodegradable TDAD polymer to be used in photoresists.

480779 Dehydration and Condensation of Furfuryl Alcohol to Light Oligomer over Molybdenum Oxide Catalysts: Effect of Loading and Reaction Time

Abstract:

The conversion of renewable biomass into bio-fuels and bio-chemicals has been investigated as a result of increasing demand for petroleum oil and increasing concern of greenhouse gas emissions. Lignocellulosic biomass (non-edible biomass) is second generation biomass composed from the residues of agriculture and forest, which prevent competition with producing biomass for food and as a feedstock as renewable biomass. The abundance of lignocellulosic materials resulted in the synthesis of large amounts of furfurals through acidic hydrolysis which is an important raw material for the synthesis of valuable chemicals. Furfuryl alcohol (FA), which is produced from furfural derived from a selective dehydration of xylose, is an attractive intermediate chemical for the production of various chemicals, such as levulinic acid and polymerized FA (PFA). PFA can be easily produced from FA monomers using a strong mineral acid, such as H_2SO_4, and can produce ordered carbons through the pyrolysis, vapor deposition and carbonization methods. Although homogeneous catalysts show a higher conversion of FA monomer to FA oligomers/ polymers without hydrothermal treatment, they should be replaced by heterogeneous catalysts for separation and recycling issues. Molybdenum Oxide (MoO_3), a metal oxide catalyst was investigated in a batch reactor system for the FA dehydration reaction under mild experimental conditions, at 100°C and ambient pressure. Five dimers (2,2â€™-Difurylmethane, 2-(2-furylmethyl)-5-methylfuran, Difurfuryl ether, 4-Furfuryl-2-pentenoic acid Î³-lactone, 5-Fufuryl-furfuryl alcohol), and two trimmers (2,5-Difurfurylfuran and 2,2â€™-(Furylmethylene)bis(5-methylfuran)) were observed through GC and GC/MS. Various catalyst loading of MoO_3 ranging from 0.05g to 1.2 g with 3.00g of FA for 6 hours were compared and the highest conversion ~36wt% was obtained with a catalysts loading of 1.2g. Two different catalyst loadings (0.3g and 1.2g) were chosen to determine the optimal reaction time. It was observed that the

FA conversion rate gradually increased when reaction time increased from 2hrs to 8hrs, while the selectivity to the observed C_9-C_{15} oligomers decreased with increase of reaction time. The highest yield to C_9-C_{15} oligomers (~21wt%) were obtained by using 1.2g MoO_3 for 8 hours. Reaction mechanism for FA dehydration over MoO3 is also proposed. FA oligomers made from the Molybdenum Oxide catalyst can be upgraded using a hydrodeoxygenation (HDO) reaction to produce alternative biofuels C_9-C_{15} in the future.

478056 Molecular Simulation of Alkaline Earth Halide Salt Solutions in Water and Organic Solvents

Abstract:

Many chemical properties are difficult or dangerous to measure experimentally, yet remain of interest for advancing efficiency in chemical processes. Molecular simulations can predict such properties a priori. Furthermore, they provide physical and chemical details on an atomistic level that cannot be resolved experimentally. In this work, we study all combinations of alkaline earth metal cations and halide anions that are experimentally water soluble using molecular simulations. We determine the densities for all salts in three solvents, namely water, methanol, and ethanol, using molecular models from the literature. Where possible, a comparison with experimental data is made, which is favorable in most cases. Additionally, we calculate the radial distribution function of the three solvent molecules around the ions, revealing a strong structuring of the solution in all cases. In subsequent studies, the employed models can be used to explore other properties, such as activity coefficients and transport properties.

478084 Relating Conductivity and Tortuosity through Modeling of Infiltrated Solid Oxide Fuel Cells

Abstract:

As the worldâ€™s energy needs continue to increase and the supply of fossil fuels continues to decrease, the demand for a clean energy source is greater than ever. One promising source of clean energy is hydrogen fuel cells. In this research we focus specifically on Solid Oxide Fuel Cells (SOFCs). SOFCs work by directly producing electricity from oxidizing a fuel. Some of the unique benefits of SOFCs include high efficiencies, low emissions, and fuel flexibility. These benefits come at the cost of high operating temperatures and cells sensitive to their surroundings. In order for the fuel cell to produce an electrical current, the electrodes must have sufficient porosity (empty space) for fuel to transport through the cell, oxygen ion conductivity to allow the oxidizing agent to react with the fuel, and electron conductivity to transport electrons created when a reaction occurs. Additionally, reactions can only take place at three phase boundary (TPB) sites, which are defined as the intersections of electrically conducting particles, oxygen ion conducting particles, and empty space. In this work, we focus on infiltrated SOFCs, which involves in extra step in the creation process of adding a material with high electron conductivity after the backbone of the cell has already been created. Infiltrated electrodes allow for greater flexibility in the design of SOFCs and have been shown to increase performance and allow for lower operating temperatures. The creation process for infiltrated SOFCs is time consuming, expensive, and involves many design parameters many of which are poorly understood. By using a model to recreate this design process, this research hopes to shape future experiments that can be run in a laboratory setting.

The model works by taking in a variety of design parameters and closely follows the anode fabrication process in order to randomly generate a cell. The fabrication process begins with the slurry formation and film creation, followed by the combustion of pore forming particles, ceramic sintering and infiltration. A variety of calculations are then performed on this cell such as effective conductivity and TPB density. This work focuses specifically on the path tortuosity of each phase within a cell and uses this information to better understand and predict the performance of each phase as well as overall cell performance. Although tortuosity has been linked with performance in past research, due to the difficulty of measuring tortuosity, studies are extremely limited and the results have varied. Highlighted results include that (1) as ceramic:infiltrate radius ratio increases in a two phase system, the ceramic tortuosity increases while pore tortuosity is unaffected, (2) as infiltrate loading (%) increases, the ceramic tortuosity is unaffected, the pore tortuosity increases, and infiltrate tortuosity decreases.

477728 A Comparative Study of Photosynthetic Unit (PSU) Models for Algal Growth Rate and Fluorescence Prediction Under Light/Dark Cycles

Abstract:

Accurate description of light-limited algal growth, especially under short light/dark (L/D) cycles, is crucial for prediction of photobioreactor performance and for optimization of operating conditions. Short light/dark cycles have been shown to increase algal light usage efficiency and productivity. A variety of widely used photosynthetic unit (PSU)-based models were evaluated to determine their ability to predict algal mean specific growth rate and mean photochemical efficiency under a range of light/dark cycle conditions. Analytical or numerical solutions were derived, where necessary, for six models for quasi-steady-state light-limited growth under L/D cycles. These models were fit to previously published experimental data for algal mean specific growth rate and mean photochemical efficiency. Subsequently the weighted sum of squared error (SSE) values, normalized sensitivities to parameter change, and corrected Akaike Information Criterion (AICc) scores were compared. The quality of the fits and the model sensitivities were used to evaluate the assumptions and relative merits of the models considered. Sensitivity analysis revealed multiple adjustable parameters which had minimal effect on model predictions or showed strong correlations with other parameters, indicating that they could potentially be fixed to improve model quality. Each model generated qualitatively similar predictions of mean specific growth rate but their relative qualities could be distinguished through differences in mean photochemical efficiency predictions. An important model assumption was the use of adjustable rather than fixed exponents for the power-law relations governing PSU inhibition and total PSU concentration with respect to light intensity. Additionally, the assumption of PSU repair to the resting state results in better fit to experimental data than does the assumption of PSU repair to the active state. The Nikolaou and Papadakis models, which had more adjustable parameters, were shown to be unnecessarily complex for accurate prediction under L/D cycle conditions. The simpler Bernardi model scored significantly better on the AICc measure and shows good potential for future use in predicting algal behavior under L/D cycles.

478169 Application of the Noh Problem to Al, Fe, W, Cu

Abstract:

Verification problems test a computer code's ability to numerically solve a problem with a known exact solution. This is useful because by correctly solving a problem with a known solution, it increases our confidence that it can solve more complex problems for which the solutions are unknown. Here we revisit the Noh verification test problem. The one-dimensional planar Noh problem is comprised of a compressible, ideal gas of negligible viscosity that is initialized with a uniform, inward velocity, impinging on a hard wall. This flow pattern creates an outgoing shock wave at the origin, and is commonly used to assess a code's ability to convert kinetic energy into internal energy. These characteristics of the Noh problem make it simple, analytically; yet it remains a very difficult numerical problem. Traditionally, the Noh problem is run using an ideal gas. Recently, however, Burnett et al, have shown that an analytical solution may also be obtained with the "stiff gas" equation of state. The stiff gas is relatively simple in form but gives a qualitatively accurate representation of many metals, while allowing condensed-phase materials to go both into compression and tension. Here we implement the stiff-gas EOS in the Noh problem to model the behavior of aluminum, copper, iron and tungsten under shocked or extreme conditions and evaluate the code's ability to follow the thermodynamic behavior of the material. These hydrodynamic verification studies were done using the LANL Eulerian code xRAGE and the LANL Lagrangian code FLAG. Numerical results are compared to the exact stiff-gas results and to the experimental shock Hugoniot, and mesh convergence studies are performed. Our results show that this classic and widely-studied verification test problem can be extended beyond the ideal gas to physically realistic materials of practical interest

478171 Genetic Algorithm Lead Discovery of Nanoparticle Structures

Abstract:

Nanoparticles are at the forefront of light and semiconductor technologies. In order to take advantage of a nanoparticle's tunable optic and electronic properties, it is important to know the 3-D structure of the nanoparticle. For example, nanoparticles of different sizes and shapes will emit varying wavelengths of light. But, because of nanoparticles' heterogenous structures, typical X-ray diffraction structure determination techniques are not able to represent the entire structure. Researchers have begun to use computational models to bridge the gap between the incomplete picture the experimental data presents about nanoparticles. These computational models do have challenges; the search space for all possible locations of each atom in the nanoparticle is on the order of 10^{139} â€" 10^{297} permutations, assuming a nanoparticle comprised of 100 atoms. This vast search space suggests the need for an intelligent algorithm to narrow the structure possiblities.Â A genetic algorithm was created to predict the 3-D structure of the nanoparticle. Genetic algorithms mimic the natural selection process in evolution. An initial population of candidates is created, and through many generations, the candidates are altered and crossed-over in order to produce a population of candidates with high fitness, or rather the best qualities. The genetic algorithm begins with bulk crystal structure predictions and alters the structure candidates through mutations and crossovers to fit Pair Distribution Function (PDF) X-ray diffraction data of a given nanoparticle. The algorithm guides each generation of candidates to have improved fits to the experimental PDF data. The dynamic computational model is used to create an advanced structural model of an experimentally observed Cadmium-Selenide nanoparticle. The algorithm is able to predict nanoparticle structures with low residuals of 0.18, meaning the model prediction is closely fitting the experimental data. With this genetic algorithm, the search space of possible structure predictions can be greatly reduced, allowing for intelligent and valuable structure guesses to be made. Modeling nanoparticle structures will lead to a deeper comprehension of structure-property relationships in nanoparticle technologies.

478178 Modeling and Control of Crystallization Kinetics

Abstract:

Crystallization kinetics is monitored from batch crystallization as a route in a phase space mapped out by the crystal mass and the total chord count. We achieve monitoring of chord count by using the Focused-beam reflectance measurement (FBRM) which works by shining a ray of light on the suspended crystals and measuring the count by the number of rays reflected back. Another instrument used to monitor crystallization is the attenuated total reflectance Fourier transform inferred (ATR-FTIR). This is typically used to monitor the solution composition and, along with temperature measurement, the supersaturation of the solution. We also use ATR-FTIR to deduce the total crystal mass via a mass balance. Manual crystallization control can be achieved by intuitively understanding the kinetics through observing the dynamics of previous crystallization runs. Through manipulation of temperature, the supersaturation of the solution is affected which in turn sets the crystallization angle. We observed that the route to reach a set target point was different between the manual manipulation and a common crystallization technique, Direct Nucleation Control. After a couple of successful attempts, we were able to develop a feedback control through understanding the crystallization dynamics. The control model proved to be more energy efficient and hits the target on the mass-count visual relatively in the same amount of time as Direct Nucleation Control. This manual control is important in testing the idea that we can understand the crystallization dynamics through the mass-count visual and establish control through learning; similar to learning how to drive.

478257 Multiplexing Techniques for High-Throughput Gas Chromatography

Abstract:

In the search for new catalysts, drugs, and organic syntheses, large libraries of substances must be analyzed. Gas chromatography (GC) is a popular analytical technique, as it is accurate, sensitive, and reproducible, although it is fairly slow (minutes to days for full analysis). This study focused on applying "multiplexing", or packaging multiple signals together into one signal, to GC. A Hadamard Transform (HdT), a variation of the Fourier Transform, was used to significantly expedite GC by deconvolving multiple overlapping sample signals. To validate the procedure, a set of GC data was simulated, and the simulated data was compared to experimentally obtained data. HdT analysis was performed on the simulated data and the deconvolved "overview chromatogram" was obtained. The original simulated data was then recalculated from the deconvolved signals, which showed that the data deviated significantly from the initial data, probably due to inaccuracies in the overview chromatogram. In the future, the deconvolution process will be improved upon and further validated. Lastly, this procedure will be applied to reactions of interest, such as dry reforming and tri-reforming.

478383 Understanding the Transferability of the Optimized Potentials for Liquid Simulations (OPLS-AA)

Abstract:

Department of Chemical Engineering and Materials Science, Wayne State University

The accuracy and transferability of the Optimized Potentials for Liquid Simulations All Atom (OPLS-AA) force field is evaluated for linear and branched isomers of alkanes up from methane to octane. Of particular interest are highly branched compounds, such as 2,2,3,3-tetramethylbutane, 2,2,3-trimethylpentane and 2,2,4-trimethylpentane. Vapor-liquid coexistence curves, vapor pressures, heats of vaporization and critical points are determined from Gibbs ensemble Monte Carlo simulations. The predictions follow experimental trends, but significant quantitative differences are observed between simulation and experiment. These results are used to provide insight into the transferability of empirical force field parameters to a broad range of molecular topologies and local bonding environments.

478480 Understanding Water Effects in Alcohol Dehydration Activity on Î³-Al2O3 Using Microkinetic Modeling

Abstract:

Increased consumption of declining petroleum resources in the production of plastics and chemicals necessitates the utilization of alternative feedstocks. Specifically, solid acid-catalyzed dehydration of biomass-derived alcohols has attracted significant attention as an excellent alternative to forming olefins, the building blocks for polymers production. Î³-Al2O3 is an active dehydration catalyst and surface reactions evolve via Lewis acid-catalyzed (concerted-E2) mechanism. Specifically, the tri-coordinated surface Al sites of Î³-Al2O3 are the strongest Lewis acid sites. However, the formation of water (dehydration product) can compete with alcohol adsorption and change the surface acidity of Î³-Al2O3. Using DFT (Density Functional Theory) calculated reaction energy profiles we constructed a single-site microkinetic model of alcohol dehydration on Î³-Al2O3. Our model addresses the activity of different-substituted alcohols, ethanol, propanol, isopropanol, and t-butanol and rationalizes the higher activity of tertiary over primary alcohols. Importantly, our model demonstrated that (competitive) water adsorption dominates on the active sites. Reaction path analysis suggests that shifting the equilibrium of surface bound species to alcohols would have a significant kinetic effect in increasing the olefin formation rates.

479624 Modeling Ultracool Brown Dwarfs with the James Webb Space Telescope

Abstract:

The James Webb Space Telescope (JWST), an international effort between NASA, European Space Agency (ESA), and Canadian Space Agency (CSA), will launch in 2018 with the intent to observe our universe from up to 1 million miles away from Earth. To enable these observations, JWST consists of a telescope, an instrument package, a spacecraft, and a sunshield. The instrument package, or the Integrated Science Instrument Module (ISIM), contains the four main instruments used to observe deep space: Near-Infrared Camera (NIRCam), Near-Infrared Spectograph (NIRSpec), Mid-Infrared Instrument (MIRI), and Fine Guidance Sensor/Near InfraRed Imager and Slitless Spectrograph (FGS/NIRISS).

The Early Release Science (ERS) program has been created in order to give the scientific community access to the JWST science observations using the four instruments aboard as early as 2019 in Cycle 1 of the mission. The purpose of doing these initial observations, which will become immediately public, is to procure representative datasets and deal with technical challenges that may arise with the varying instrumental modes on JWST. The main goals to be accomplished for brown dwarfs through the ERS program include: a possible 20-30 hour proposal, obtaining mid-infrared spectroscopy and photometry of some representative brown dwarfs with a range of temperatures and surface gravities, enabling future larger proposals, test models, and to inform the interpretation of expensive exoplanet spectra. With these goals in mind, we are currently investing brown dwarfs of interest in the correct regions of the sky and modeling what we could expect to see using the MIRI IR Imaging Exposure Time Calculator (ETC).

479684 Advanced Model Predictive Feedforward/Feedback Control of a Tablet Press

Abstract:

In continuous pharmaceutical manufacturing, real time precise control of critical quality attributes (CQAs) is necessary for Quality by Design (QbD) based manufacturing and real time release (RTR) with minimum consumption of time, space and resources. Pharmaceutical tablets can be produced through different routes with a common tablet press unit operation always placed at the end of the manufacturing process. Therefore, the tablet press is a crucial unit operation directly influencing the CQAs irrespective of manufacturing routes. Despite this, little attention has been paid to the development of an advanced efficient control system for the tablet press. Process modeling can be used as an efficient virtual experimentation tool to design, compare and evaluate different control systems. We developed a model in Simulink (Mathworks) that includes two master control loops for tablet weight and hardness and a slave feedback loop controlling the compaction force applied to each tablet. We examined the performance of different control strategies based on Proportional Integral Derivative (PID) control and Model Predictive Control (MPC), as well as feedforward/feedback control. We found that a hybrid MPC-PID control strategy outperforms the PID only control strategy. We also observed that the addition of feedforward control further improves the performance of hybrid MPC-PID control strategy.

479274 Modeling and Simulation of Pollution Control Units for Improved Sustainability: A Styrene Production from Ethylbenzene Case Study

Abstract:

The United States Environmental Protection Agency (U.S. EPA) has long advocated for, and mandated, regulations on chemical processes to prevent and reduce the life cycle impact the processes have on the environment, as processes in the past have created pollution to the extent that it has affected the environment and climate of the world around us. In order to cope with the growing problem of pollution by the release of various chemicals into the environment, it is necessary to include unconventional pieces of equipment to purify and clean waste streams when designing chemical plants. Pollution control unit simulators are needed to be included when assessing, improving, and optimizing the process performance in terms of sustainability. Flare, as one important and typical air pollution control technology, can treat most of volatile organic chemicals (VOC) streams with different mass and energy flows. In this project, a flare unit model is developed in Microsoft Excel to calculate the equipment life cycle inventory (LCI). This LCI consists of mass and energy balances, equipment sizing, and utility demand for a chemical process represented by a CHEMCAD simulation.

In order to test the unit, a chemical process model for the manufacture of styrene from ethylbenzene is developed in CHEMCAD. This CHEMCAD simulation is then connected via data mapping to the Microsoft Excel spreadsheet containing the flare unit model. Specifically, the output from the gaseous waste stream of the styrene production process is connected to the flare unit, where the waste products are treated to minimize the harmful outputs, thus increasing the process sustainability. In this presentation, different scenarios associated with the connection of the flare unit to the styrene production process will be discussed towards a more sustainable process operation.

479288 Modeling, Simulation, and Control of a Supercritical Coal-Fired Power Plant for Smart Grid Applications

Abstract:

Modeling, Simulation and Control of a Supercritical Coal-fired Power Plant for Smart Grid Applications

Jacob Douglas, Xin He, and Fernando V. Lima

Submitted to Topical Area: Computing and Process Control

Modern smart grid programs rely on the combustion reaction of fossil fuels such as coal and natural gas to supplement energy produced by renewable resources. These programs utilize several different sources of energy such as wind, solar, and fossil fuels. In a smart grid program, the power required from fossil fuel power plants has to change quickly in response to the available renewable energy. Therefore a fossil fuel power plant has to be able to ramp up or ramp down power generation instantaneously to account for the intermittent deficits of renewable energy. Several different power plant designs have been proposed to try to solve the problems associated with the implementation of a smart grid program. One of these designs is the supercritical coal-fired power plant. An advantage of this design is that the power plant runs at a constant pressure. Thus, if the temperature entering the turbine cycle is kept constant as well, the power produced by the design can be cycled by changing the flow rates of the water and coal entering the process. With this design, the power produced from the combustion of hydrocarbons can be manipulated quickly to account for a change in the energy produced by renewable techniques.

In this presentation, the conditions of turbines and a boiler used in the supercritical coal-fired power plant design are manipulated by using proportional-integral-derivative (PID) controllers to show how this plant design could be incorporated into a smart grid program. In the control scheme for the developed power plant model, the pressure and temperature of the steam entering the turbines are kept constant at specified set points. The implemented control scheme allows for the inlet conditions of the solid coal and water flow rates to be cycled in order to obtain a desired power output. Preliminary results indicate that a cycling range of ± 25% is achievable by the current power plant design. These results show that using this controller design, it is possible to make changes to the plant model to accommodate energy produced by renewable sources.

479728 Chemical Game Theory: An Approach for Modeling Strategic Decisions

Abstract:

Sustainability has both technical and human decision-making factors. In fields such as economics and political science, human decision-making is frequently analyzed using game theory. However, the classical model of game theory has several shortcomings. The purpose of this poster is to propose an alternative method for solving games, in which players' strategies are treated as reactant molecules and equilibrium decisions are evaluated using Gibbsian thermodynamics. We call this alternative method "chemical game theory." Chemical game theory removes shortcomings of classical game theory by including entropy in the game solution and incorporating player biases, outside enforcer agents, and payoff magnitudes. This poster will focus on the Prisoner's Dilemma game, a common 2x2 game used to represent Tragedy of the Commons problems. It will illustrate the distinctions between the chemical and classical solutions, and apply chemical game theory to topics of climate change, the state budget process, and the food-water-energy dilemma.

479906 Design and Construction of Pilot Scale Tangential Flow Filtration Unit

Abstract:

Design and Construction of Pilot Scale Tangential Flow Filtration Unit

Purpose:

The goal of this study is to design and build a pilot scale tangential flow filtration (TFF) unit to provide a new experiment for chemical engineering students to perform in their unit operations lab. The unit can be used as both a mass transfer experiment and a process control experiment. In mass transfer mode the student will gain hands-on experience with TFF. In process control mode, students will be challenged to tune two coupled control loops to regulate the transmembrane pressure and flow rate simultaneously.

Equipment:

The unit utilizes a 1.14 m^2 10 kDa ultracel membrane filter. The system features two 55 liter holding tanks a 5 horsepower pump controlled with a variable frequency drive and a number of manual on/off valves that allow the user to create different pathways for the system. Three magnetic flow meters are used to measure permeate and retentate streams. Three pressure transmitters are used to measure the transmembrane pressure using the equation: $TMP = (P_{feed} + P_{return})/2 - P_{permeate}$.

The TFF unit is connected to and operated via a distributed control system (DeltaV). A local operating station is located next to the unit. A graphical user interface (GUI) provides for intuitive operation in mass transfer mode and controller tuning in process control mode. All process data are recorded and archived in a data historian.

Operation:

In a typical mass transfer operation mode the feed solution is pumped out of one of the tanks into the membrane. The permeate is collected in the second tank while the retentate is returned to the feed tank. The user can set the TMP and permeate flow rate simultaneously. The TMP is controlled by a control valve in the retentate line. The permeate flow rate is controlled by the pump speed.

In process control mode the operation is similar to the operation in mass transfer mode but the membrane module is bypassed and the flow resistances associated with the membrane are simulated with globe valves.

With the valve and membrane pathways on the TFF unit, students will be able to perform two different experiments in lab. With the valve system, students can understand the effects of coupled control loops. With the ultracel membrane, students can perform bioseparation experiments at various operating conditions and quantify the performance of the membrane. The largest risk with this project is exceeding the pressure limit of the membrane. A safety interlock needed to be developed in order to prevent the TMP from reaching a level high enough to damage the membrane. The interlock was a section of code added to the pump control algorithm and was designed to stop the pump immediately if the TMP exceeded a predetermined value. The safety interlock was tested several times and successfully stopped the pump when a spike in TMP was detected.

480344 Developing a Molecular Model to Predict Equations of States from Computational Quantum Chemistry and Machine Learning

Abstract:

Computational models can allow engineers to design advanced transportation engines that operate at significantly high pressures. In order to improve fuel-air mixing in the engines and increase efficiency, chemical engineers need to understand the thermodynamic properties of the mixture under these conditions. An equation of state (EOS), which represents a mathematical relation of volume of a mixture to temperature and pressure, offers a valuable tool for understanding important properties of mixtures such as vapor-liquid equilibria and other thermodynamic quantities. For example, important fluid properties such as compressibility factor, flow density and critical conditions of hydrocarbon mixtures can be derived from the EOS to predict the condensation of gas mixtures in pipelines.

Given that EOS is valuable in deriving essential thermodynamic properties, the challenge of current technology is the ability to predict the EOS for hydrocarbon mixtures at these high pressures with better accuracy. The current method of empirically determining accurate EOS through laboratory measurements is problematic because it requires too much time and many experiments. In addition, the number of transportation fuels is constantly increasing. Biofuels, such as alcohols, ethers and ketones also behave differently from more traditional petroleum-derived alkanes due to their chemical bonding and molecular structures. Consequently, there is an increasing demand for EOS for new mixtures and the ability to accurately predict the EOS for any mixture of interest. Therefore, designing a better approach for applying EOS to an arbitrary mixture would represent a significant scientific achievement that would result in major breakthroughs in designing advanced engines.

The goal of this project is to examine the practicability of predicting equations of state, focusing on a specific class of EOS known as the virial equation of state. Theoretically, important properties of mixtures can be accurately determined from the virial EOS. However, the lack of accurate parameters limits our ability to accurately predict

intermolecular interactions. I will apply computational quantum chemistry to produce these parameters. In this project, I will employ a two-step method in this project. First, I will identify the coordinate system and the degrees of freedom of the system. This process will produce a mathematical description of a multi-dimensional function of intermolecular potential. Second, the Electronic Structure Method will be employed to sample values of intermolecular potential to solve for the second virial coefficient. The Monte Carlo Integration method can be used to estimate integrals that cannot be evaluated analytically, such as the intermolecular potential function in this model. Further, I can apply my method to study the contribution of chemical functional groups in more complicated molecular system and calculate the third virial coefficient. By the end of the project, I will have developed a high-performance computational model to compute pairwise potential, enabling chemical engineers to simulate and design their processes with greater efficiency.

479936 Molecular Dynamics Determination of the Effects of Correlated Hopping on the Kinetic Monte Carlo Epitaxial Growth Patterns on Ni(111) Versus Ni(100)

Abstract:

In this work we use molecular dynamics (MD) to systematically examine and measure the probabilities of correlated adatom motion for Ni adatom growth patterns on base Ni(111) versus base Ni(100) surfaces. The interaction energies are determined by classical-potential total-energy calculations. We examine adatom size effects and temperature dependence over a 100K to 2000K range. Our results indicate that over this range of experimentally interesting temperatures, correlated motion on the Ni(111) surface is indeed very significant, even at low temperatures, while on the Ni(100) surface it is insignificant even at the higher temperatures.

480572 A Monte Carlo Model of Nanomaterial Toxicity

Abstract:

In the field of nanomedicne, tumor imaging agents such as quantum dots are introduced into the body by injecting metamaterials into the blood stream. In many cases, however, the kidney and liver will collect these nanomaterials and aggregate them preventing them from being removed from the body and increasing nanomaterial toxicity.Â On the other hand, a vast majority of nanomaterials are removed through the kidneys; therefore, if we can prevent the nanoparticles that collect in the kidney from aggregating we can remove them from the body quickly and dramatically reduce their cytotoxicity. Previous strategies for optimizing the removal of nanoparticles have focused on â€œsize optimizationâ€ or restricting the size of the particles to be less than 5.5nm. However, recent studies indicate that toxicity and removability depend more on seemingly irrelevant factors like shape, charge, bonding strength and a multitude of other variables. Although there is a lot of experimental data on nanomaterial toxicity, the sporadic dispersion of toxicity levels that depend on a variety of variables has created a high demand for an all-inclusive model to predict the toxicity of novel nanomaterials without expensive testing. This is epically important due to the nature of nanomaterials research, which frequently experiments with new compounds and structures.

Ideally, density functional theory can provide the most accurate model of nanomaterial toxicity, but the computational expense of this modeling technique encourages us to simplify the simulation using molecular dynamics. We can accomplish this by using density functional theory to characterize the chemical interactions between the nanomaterials and the body but then using molecular dynamics to account for size and shape. The randomness of the interactions between the nanomaterials and the cell membranes will be best captured though a Monte Carlo model.

Through using a Monte Carlo model of hard-spheres we have been able to show that nanomaterials are less likely to aggregate near the membranes of kidney cells. The implications of this experiment will act as a guide for developing clinically applicable nanomaterials and be included in the overall toxicity model that will unleash the medical

capabilities of nanotechnology.

480738 Software Package for Non-Uniform Sampling

Abstract:

Alternative Sampling Densities in Non-uniform Sampling

D. Levi Craft*, David Rovnyak*, Adam Schuyler^

* Bucknell University, Department of Chemistry, Lewisburg PA 17837

^ University of Connecticut Health Center, Biophysics Department of Molecular Biology and Biophysics, Farmington CT 06030

Non-uniform sampling (NUS) often means acquiring time domain samples non-consecutively following a sampling density. Robust spectral estimation techniques have been developed to produce high fidelity frequency spectra from NUS NMR data, where it is appropriate to view the samples as constraints on line shapes. Exponential weighted sampling paired with maximum entropy reconstruction (MaxEnt) has been shown through extensive empirical work to yield frequency spectra highly comparable to those that would be obtained by FFT of uniform data.

Line shapes obtained by NUS-MaxEnt may experience a very weak, but non-negligible broadening only near the peak base when employing exponential weighting matched to the signal decay (figure below). However, this problem can be resolved by sinusoidal weighting of an exponential signal, which has the same intrinsic sensitivity as matched exponential NUS, but uses more samples in the regime 0.5-2.0 T2, and does not suffer broadening of peak bases (Figure 1).

Motivated by this finding, we set out to further improve alternatives to exponential densities. For example, a sampling density based on a quadratic function has slightly higher intrinsic SNR than sinusoidal and matched exponential NUS,

while still preserving the shapes of peak bases (Figure 1).

Exponential NUS can be biased to even shorter times in order to obtain intrinsic SNR enhancements over two-fold compared to uniform sampling,1ab but the extra SNR comes at a cost. The line width at half height is not perturbed by matched exponential NUS, but is broadened if the NUS is biased further.1b,2 We show that much of the broadening incurred by biased exponential NUS can be mitigated by using alternative densities, such as a squared sinusoidal function that has nearly the same intrinsic SNR, but acquires more samples after 0.5T2 to better constrain the line shape (Figure 2).

Finally, we recently developed a simple algorithm for producing properly gapped, sinusoidally weighted NUS schedules that retain randomness.2 For this work, we extended the algorithm to more sampling densities, and also implemented a new set of deterministic statistical NUS schedules. We will show that MaxEnt-NUS spectra obtained from either the deterministic or the gapped-randomly-weighted schedules were essentially indistinguishable, validating both approaches and giving experimentalists a larger choice of robust NUS schedules.

I will contribute a set of computer programs to a larger software platform hosted at the University of Connecticut Health Center. The aim of my work is to improve the accessibility and effectiveness of an emerging technology known as non-uniform sampling (NUS) for nuclear magnetic resonance (NMR) data collection. My work will help to automate the use of NUS by non-expert users and help to facilitate the sharing and communication of data acquired by NUS.

By implementing sophisticated schedule generators and schedule metrics in programs that are accessible to non-expert users, I hope to develop clear and helpful demonstrations of the benefits of using NUS in NMR data acquisition.

1. (a) Magn. Reson. Chem. V49, 483-491, 2011. (b) J. Phys. Chem. B 116(25): 7416-7427 (2012).

1. J. Biomol. NMR, V58: 303-314, 2014.â€"

480607 Molecular Dynamics Modeling of Water and NaCl Transport in Sulfonated Poly(arylene ether sulfone) Desalination Membranes

Abstract:

Highly charged polymer membranes have been examined as possible replacements for polyamide membranes in desalination applications. While these membranes may reduce pre and post-processing costs, optimization of separation performance is challenging, due in part to an insufficient understanding of the link between atomic scale chemistry and permeate transport. In this work, a series of 140,000 atom molecular dynamics models are used to study NaCl and water transport through sulfonated poly(arylene ether sulfone) desalination membranes. A state-dependent analysis framework is developed, using custom analysis tools to identify dominant transport mechanisms within the membrane. Salt bridging and Na-sulfonate cluster formation play a key role in mediating Na interactions with the membrane, and increase the binding time of Na to fixed charges on the polymer backbone. Additionally, we find that the bulk diffusion of water, Na, and Cl is subdiffusive within the membrane for time intervals shorter than 1.5 ns. Position-dependent mean squared displacement calculations suggest this behavior could result from intermittent binding interactions of ions with sulfonate groups and slow interpore transport. This result highlights a disconnect between short-timescale molecular interactions and the ultimate bulk diffusion of membrane permeates. Further numerical modeling and more extensive characterization of the effects of membrane morphology are necessary to connect atomistic membrane interactions to the bulk transport of permeates.

478014 Encapsulation of Microbes and Slow Release Substrate for Bioremediation of Chlorinated Solvents

Abstract:

Encapsulation of microorganisms in a gel matrix offers several advantages in groundwater remediation, such as high biomass and resistance to environmental stressors. Other compounds, such as primary substrates, can also be captured in the gel matrix, providing more favorable conditions for the microbes. The present research considers the use of tetra-n-butyl orthosilicate (TBOS) as a slow releasing primary substrate for microorganisms encapsulated in a variety of gels. TBOS can be hydrolyzed to create four butanol molecules, which can be used as a primary substrate by *Rhodococcus rhodochrous*.

Calcium alginate, agar, agarose, and polyvinyl alcohol (PVA) gels were each used to encapsulated *R. rhodochrous* and determine the rate of substrate transformation. Each gel was also tested for feasibility of encapsulation with TBOS. Batch tests with *R. rhodochrous* and TBOS encapusulated in agarose beads were performed to test the ability of the encapsulated microbe to metabolize repeated injections of a mixture of 1,4-dioxane, 1,1-dichloroethene, and cis-dichloroethene.

Cells encapsulated in calcium alginate, agar, and agarose gels had similar rates of substrate transformation to suspended cells. Cells encapsulated in PVA had a lower rate of substrate transformation due to reduced diffusion of oxygen and substrates into the gel. It was possible to encapsulate TBOS into agar, agarose, and PVA gels. Encapsulation of TBOS into calcium alginate gels was possible with use of PVA to reduce interfacial tension between the alginate solution and TBOS. TBOS was also shown to be a viable slow release substrate in the long-term batch tests, as multiple injections of the chlorinated solvent mixture were metabolized without further addition of substrate, although it was less successful than the control reactors with suspended cells. Future work will aim to reduce the size of the gel beads from 5 mm to less than 100 Î¼m to allow for transport of beads within an aquifer. Tests of chlorinated solvent transformation by encapsulated cells and TBOS will be continued in continuous-flow column reactors.

Keywords: Chlorinated solvent, bioremediation, Rhodococcus rhodochrous, microbial transformation, immobilization, encapusluation

478118 Effect of Surfactants on Cloud Condensation Nuclei

Abstract:

Aerosols are present everywhere. They are formed simply by diffusion of particles into the air from surfaces of materials all around us. Inevitably, these particles will end up in the atmosphere, and many of these particles will end up acting as Cloud Condensation Nuclei (CCN). A CCN particle forms a surface on which water molecules accumulate. These CCN particles might be very small but can hold a very large amount of water. In the atmosphere, these particles are of great importance, because as the name might suggest, clouds form because of them. An ability to manipulate the uptake of water by these particles might be of great value in relatively new ideas like cloud seeding (manipulating precipitation) or global climate modeling, which could help us understand more about climate sensitivity.

This research project was conducted during the years 2015-2016 to investigate the effect of surfactants on the uptake of water by dry aerosols. For this purpose, a dilute solution of a solvent (typically 1 gram per liter) in deionized water was taken in a syringe that was connected to a device allowing a constant flow of the given solution. It would then be turned into an aerosol by the â€œatomizerâ€, a device that blasts the very small stream of the given solution with a large amount of air. Then the resultant aerosol flowed through a series of dryers that removed all the water. The final product was passed through a particle counter and a Cloud Condensation Nuclei Counter (CCNC). The particle counter would count the number of particles flowing through the tubes and the CCNC acted to pass the aerosol through a supersaturated humid environment and to measure the number of particles that would take up water and grow in size, and the size of those particles.

The idea behind this project was to see whether the uptake of water by the particles could be manipulated by adding surfactants. Surfactants (short for surface-active agents) are substances that affect the surface tension of water (or any solvent that they are dissolved in). A decrease in surface tension was theorized to be a factor in the amount of water that could accumulate around a particle, and various surfactants were tested throughout the course of the research project to see their effects on the growth of particles.

478220 Tetraethylenepentamine Functionalized Ti-MCM-36 for Carbon Dioxide Capture

Abstract:

Solid adsorbents for carbon dioxide capture have many benefits over traditional liquid amine absorbents, such as low volatility, lower corrosivity, and greener operation requirements. However, these materials can also suffer from lower capacities for multi-cycle use, with any sorbents showing either low capacity that is consistent over multiple cycles or high initial capacities that degrade very quickly upon cyclic use. [1,2] One-way to improve the performance of solid adsorbents it to post-synthetically modify them with amine groups through either physical impregnation or covalent grafting. This modification has tradeoffs and will often slow adsorption kinetics due to the blocking of diffusion channels with loaded amine groups. [3]Previously we studied the sorption kinetics of 1-dimensial pore structures, determining that excessive amine loading can degrade the carbon capture performance of functionalized adsorbents. We have begun investigating the functionalization of materials with 2-dimensional pore structures using Ti-MCM-36 in order to figure out the relations of the pore structure of support materials and the adsorption kinetics of supported amine adsorbents. Ti-MCM-36 is a derivative of the well-known pillared zeolitic material, MCM-36, that consists of silicate layers pillared with TiO_2 columns. We have shown that Ti-MCM-36 has a higher affinity for CO_2 adsorption than Si-MCM-36, with high capacities despite lower surface areas. In this work, we have attempted to further improve the materials by the introduction of TEPA (tetraethylenepentamine).

(1) Andirova, D.; Cogswell, C. F.; Lei, Y.; Choi, S. Effect of the Structural Constituents of Metal Organic Frameworks on Carbon Dioxide Capture. ***Microporous Mesoporous Mater.*** **2016**, ***219***, 276–305.

(2) Andirova, D.; Lei, Y.; Zhao, X.; Choi, S. Functionalization of Metal-Organic Frameworks for Enhanced Stability under Humid Carbon Dioxide Capture Conditions. ***ChemSusChem*** **2015**, ***8***, 3405–3409.

(3) Cogswell, C. F.; Jiang, H.; Ramberger, J.; Accetta, D.; Willey, R. J.; Choi, S. Effect of Pore Structure on CO Adsorption Characteristics of Aminopolymer Impregnated MCM-36. ***Langmuir*** **2015**, ***31***, 4534–4541.

478243 Generation and Use of a Pure Titanium Pillared MCM-36 Structure As a High Efficiency Carbon Dioxide Capture Platform

Abstract:

The drawbacks of liquid amine streams, the current industry standard for carbon dioxide capture, has fueled research into solid layered absorbent materials. Zeolitic lamellar materials have attracted interest as a platform for carbon dioxide sequestration due to the possibility of post-synthetic functionalization, such as swelling, pillaring, or exfoliating. Here, we introduce a novel functionalization of MCM-22-P precursor with titanium oxides, which upon intercalation, hydrolysis, and calcination generates a titanium pillared derivative we term Ti-MCM-36. At room temperature, this material demonstrates superior capture capacities compared to MCM-36 pillared with silica (1.35 mmol/g for Ti, 1.02 mmol/g for Si), despite having less surface area (334 m^2/g for Ti, 506 m^2/g for Si). Ti-MCM-36 demonstrates full regeneration capability through multiple cycles of adsorption and desorption, and has decreasing capture capacity with increasing temperature. These together suggest that carbon dioxide adsorbs onto the surface of Ti-MCM-36 through a physisoption mechanism. Preliminary studies on Ti-MCM-36 in humid conditions suggests that the titanium derivative has decreased hydrophilicity as compared to silica pillared MCM. This may allow the Ti-MCM-36 to have superior carbon dioxide capture in humid environments, such as flue gas streams. The improved properties of this material provides a strong foundation for future investigation.

478749 Watershed and Lake Bio-Geochemistry: Movement of Phosphorus and Sediments from Forested Areas into Water Bodies

Abstract:

Eutrophication is known as an excess of nutrients (mainly nitrogen and phosphorus) in water bodies, specially rivers and lakes, due to runoff from lands which causes an incontrollable growth of algae that can cause animal death because of food, oxygen and habitat shortage. Lake Champlain, in the state of Vermont, has been experiencing an algae bloom problem through the past years. Many factors are believed responsible for the bloom, like for example agricultural practices, soil erosion and climate change. It was determined that phosphorus (P) is one of the main nutrients responsible for the algae blooms, especially in summer where the temperature tends to increase; also is the one factor that can actually be mitigated. This research was focused in Missisquoi Bay, which constitutes only 0.8% of the total volume of the lake. In the Missisquoi Bay three sites were studied, Ross Brook and Kings Hill Brook. Each one of these sites with different surroundings that contribute to the phosphorus (P) concentrations in the soils and water streams. Ross Brook has a maple sugaring farm and a livestock area where they had access to the stream and Kings Hill has a maple sugaring farm nearby and it is believed to be a logging site. Each site was water sampled by an ISCO during each big storm. Data was downloaded where levels, samples, temperature and turbidity values were obtained. After recollecting data, all the samples were taken to the chemistry laboratory analyze and quantify total phosphorus (TP), soluble reactive phosphorus (SRP) and total suspended solids (TSS). Further research would try to demonstrate that high phosphorus concentrations in the lake result mainly from agricultural activities rather than forested areas; given that agricultural activities are common in the state.

478561 Analysis of the Impact of the Environmental Noise in the Academic University Communities: Study of Case

Abstract:

The noise as the third environmental pollutant and recent indicator of the quality of life in the cities is the cause of adverse effects in the atmosphere, affecting the health and well-being of the people, which in the long term you can generate a progressive degradation of the health of those living in the areas most affected by noise pollution. In the university communities its effect is particularly relevant because it affects non-clinical manner causing deconcentration, stress and low productivity, which affects the welfare of those who are daily in the university campuses. This situation has prompted this research whose goal is to analyze the impact of environmental noise in the university academic communities, taking as a case study of the Faculty of Mines of National University- Medellin campus. The methodology is applied and descriptive field of cross-cutting nature, by means of a methodological triangulation that incorporates review of secondary sources of information, the application of a quantitative methodological design. And the analysis of the noisy data reported by the Laboratory of air quality (CALAIRE) on your workstation the faculty of mines of National University. Among the results, it notes that between 92% and 96% of the students, independent of their level of hearing sensitivity, expressed that did not take any action to mitigate the noise. Finally, among the non-clinical disorders more common product of the noise in the students is the stress, anxiety and decrease in the concentration.

479076 Direct Measurement of Aerosol Surface Tension As a Function of Humidity

Abstract:

Aerosols are small solid particles suspended in a gas. Atmospheric aerosols are particles suspended in the atmosphere. Atmospheric aerosols are known to directly influence the global climate [Boucher et al., 2013]. Additionally, these aerosols can indirectly affect climate when they form cloud seeds [Seinfeld and Pandis, 2006]. Over time, these cloud seeds, also called cloud condensation nuclei (CCN), aggregate and grow to form clouds which have immense implications for our atmosphere and weather patterns.

Particle surface tension is a necessary parameter to model this cloud formation process. It is difficult to collect enough of the aerosol to use common surface tension instruments. This research focuses on directly measuring atmospherically relevant aerosol surface tension using the atomic force microscope (AFM). This technique allows surface tension measurements on nano to pico liter volumes of collected aerosol. The method oxidizes volatile organic compounds and then uses particle impaction to collect a film of the aerosols on a small metal puck and then the AFM probes the puck with specialty tips.

This method uses specifically designed AFM tips to measure forces involved with submersing and withdrawing the tip in the collected sample. Initial testing on liquids with known standard surface tension values verified the validity of the method. Further testing proved that oxidized Î± -pinene particles formed in dry conditions have a surface tension of that similar to the pure liquid, while oxidized Î± -pinene particles formed in more humid conditions have a significantly higher surface tension than that of the pure Î± -pinene liquid [Hritz et al, 2016]. Further experiments are being conducted in an attempt to clearly uncover the relationship between surface tension of atmospherically relevant aerosols and the humidity at which they were formed.

References:

Boucher, O., Randall, D., Artaxo, P., Bretherton, C., Feingold, G., Forster, P., Kerminen, V. M., Kondo, Y., Liao, H., Lohmann, U., Racsch, P., Satheesh, S. K., Sherwood, S., Stevens, B. and Zhang, X. Y.: Clouds and Aerosols, in : Climate Change 2013: The Physical Science Basis. Contribution of Working Group I to the Fifth Assessment Report of the Intergovernmental Panel on Climate Change, Cambridge University Press, Cambridge, United Kingdom, 571-657, 2013.

Hritz, A. D., Raymond, T. M., and Dutcher, D. D.: A method for the direct measurement of surface tension of collected atmospherically relevant aerosol particles using atomic force microscopy, Atmos. Chem. Phys., 16, 9761-9769, doi:10.5194/acp-16-9761- 2016, 2016.

Seinfeld, J. H. and Pandis, S. N.: Atmospheric Chemistry and Physics, Wiley, Hoboken, NJ, 2006.

479514 Influence of Physical Microstructure and Surface Properties on the Drying Rate of Water from Simulated Soil Micromodels

Abstract:

Bacteria have been shown to regulate moisture content in soil through the secretion of extracellular polysaccharides (EPS), a polymeric substance whose properties vary with bacterial species and environmental conditions. The presence of small amounts of EPS have been shown to significantly increase water retention in the rhizosphere soil compared with nearby bulk soil that is lacking EPS. The mechanism whereby EPS promotes increased moisture content may include (i) water being held within the swelling polymer matrix during wet conditions and remaining hydrated during dry soil conditions, (ii) promotion of soil particle aggregation which increases the capillary forces on water in small intra-particle soil pores, and (iii) altered soil surface properties, including soil water repellency. Studies conducted using bulk soil may struggle to differentiate among these mechanisms. Here we employed emulated soil micromodels with systematically controlled EPS effects such as soil structure and surface properties to better elucidate the mechanism for EPS-mediated water retention in the rhizosphere. Micromodels were fabricated with aggregated or non-aggregated structures by modifying the spatial distribution of an identical sandy loam-sized particle arrangement between micromodel designs. Micromodels of each design were then surface-treated to exhibit properties of rhizosphere versus bulk soil. Micromodels were initially fully saturated and then allowed to dry in a temperature and humidity controlled environment. Results showed that micromodels with more hydrophobic surfaces took four times longer to completely dry compared to micromodels with less hydrophobic surfaces. Furthermore, microstructure geometry was found to control the spatial distribution of moisture within soil micromodels. These results illustrate the mechanisms whereby soil structure and surface properties influences the drying process in the rhizosphere. Future studies should focus on the effects that different chemical constituents found in bacterial secretions may have on moisture retention using this emulated soil micromodel approach.

479543 Photocatalytic Oxidation-Based Water Treatment Using Nano-Structured Tio 2 Electrode: Co-Catalyst Vs. Non-Ferric Fenton

Abstract:

Advanced oxidation processes (AOPs) are significant in that they can overcome kinetic limitations of conventional oxidation processes in the treatment of organic contaminants. Doped TiO_2 (titania), and un-doped titania in conjunction with an electron-donor such as H_2O_2, can be used as electrodes in photocatalytic AOPs. In this study, the dye methyl orange (MO) was used as an example of the target organic contaminant. Photocatalytic degradation of the dye was studied initially using as-produced and palladium-doped titania photo-electrodes prepared using the anodic oxidation method, with nanotube (NT) and nanorod (NR) structures. Palladium was doped on the electrodes using a $PdCl_2$ precursor with illumination under mercury lamp. SEM and XRD were used for surface analysis and UV-Vis for monitoring photocatalytic degradation. Kinetic parameters were obtained using the Langmuir-Henshelwood model. It was determined that the rate of MO degradation was highest for the Pd-TNT electrode and 5 times slower for the Pd-TNR and the as-produced un-doped TNT electrode. The as-produced un-doped TNT electrode was then employed (coupled with an inert electrode, without any external bias) along with varying concentrations of H_2O_2, for MO degradation. Complete degradation of the dye was observed at a much faster rate than that for heterogeneous photocatalytic oxidation of the same contaminant using the Pd-TNT electrode.

478775 Variability of Electronic Cigarette Particle Size Distributions and the Effects of Background Air

Abstract:

In the last few years, electronic cigarettes have been increasing in popularity, especially with younger people and people attempting to quit smoking. However, little is known about the various factors that affect the effluent from these devices. E-cigarettes are powered by a battery that heats up filament which in turn heats up a wick and converts the e-liquid into vapor that is then inhaled by the user. The e-liquids are composed of propylene glycol, glycerin, varying flavor agents, and a range of nicotine concentrations. The main goals of this research project are to identify factors which affect the size distribution of particles in the effluent. The factors we are examining include the relative humidity of the feed air, the concentration of particles in the feed air and in the environment that effluent immediately mixes with as well as the heating voltage and the composition of the e-liquid.

The nicotine content in the gas phase increased with an increase in power to the e-cigarette, but in general the majority of the nicotine was found in the particulate phase. The common range for e-cigarette operation today is between three and five volts, and there was a greater nicotine content that was found when operating the e-cigarette at the higher end voltage. This research produced data that showed the different varieties of e-cigarettes produced bimodal distributions of aerosol particle sizes, which is associated with the different boiling points of the different liquid components.

479864 Particle-Induced Gamma-ray Emission (PIGE) as a Novel Detection Method for PFAS in Personal Care Products

Abstract:

Per- and polyfluoroalkyl substances (PFASs) have been widely used in industry for their well-known water-repellent qualities and surfactant properties. Because of the environmental persistence of PFASs, their ability to bioaccumulate, and their suspected human toxicity, new methods to identify these chemicals at trace levels in personal care products (such as cosmetics) are needed. Particle induced gamma-ray emission (PIGE) spectroscopy is an established ion beam analysis technique that has been used to quantitatively measure light elements in diverse target materials including water, soil, fabric, and paper samples. Since the total fluorine measured by PIGE in other sampling media has correlated well with the organofluorine content, PIGE was further employed to act as a screening process for PFASs in personal care products such as mascaras, skin and lip care products, eye shadows, eyeliners, and make-up primers. Subsequent experiments with cosmetic samples from a range of brands demonstrate PIGE to be an inexpensive, rapid, and non-destructive method for total fluorine analysis, which can be adapted to detect environmentally relevant PFASs in care products.

479919 Measurement & Flux Modeling of Lake Michigan's Polychlorinated Biphenyl (PCB) Content Near Chicago

Abstract:

Polychlorinated biphenyls (PCBs) exist in various matrices such as air, sediment, serum, soil, tissue, and water and pose a risk of toxic effects in humans and ecosystems; proclivity to bioaccumulate and biomagnify. In addition to being known carcinogens, PCBs also exhibit effects in the immune, reproductive, neurological, and endocrine systems. Dietary exposure and inhalation have been shown to be the primary pathways of human exposure. This study hypothesized that Chicago is an ongoing source of airborne PCBs to Lake Michigan. To address this, the overall strategy employed by this study was to simultaneously sample air and water to measure PCB concentrations and determine the fate of PCBs in southern Lake Michigan.

This study has successfully completed a detailed analysis of the quantity and transport of southern Lake Michigan's PCB content; developing a detailed mathematical flux model in the process. The model predicts the air-water exchange though application of physical-chemical properties of each PCB congener, meteorological data, and environmental sampling. The air-water exchange was modeled, utilizing MATLAB®R2015a, with combination of the following: Whitman's two-film resistance model, Deacon's boundary model, three-phase partitioning model, and two parameter Weibull distribution.

Samples were collected, 20 in total, aboard the U.S. Environmental Protection Agency's R/V *Lake Guardian*, in September of 2010; 5 km off the coast of Chicago. The environmental samples were processed employing an Accelerated Solvent Extraction (ASE) System 300 and analyzed utilizing gas chromatography with tandem mass spectrometry.

Gaseous ΣPCB concentrations ranged from 0.19 to 1.1 ng m^{-3} with an average of 0.75 ± 0.3 ng m^{-3}.

Furthermore, dissolved ΣPCB concentrations ranged from 150 to 170 ng m^{-3} with an average of 160 ± 13 ng m^{-3}. Based on the Aroclor history of the area, the air and water exhibited expected trends: combination of Aroclor 1242, 1248, and 1254. For the entire period of study, the congener specific net flux ranged from -27 to 18 ng m^{-2} day^{-1}, PCB107+124 and PCB11, respectively; net ΣPCB fluxes were determined to range from -320 to 84 ng m^{-2} day^{-1} with an average of -160 ng m^{-2} day^{-1}. Positive and negative fluxes indicate volatilization and absorption, respectively. Despite the limited study period, this study displays strong evidence that Chicago is a source of PCBs, primarily mid to higher chlorinated congeners, to Lake Michigan. Furthermore, Lake Michigan was found to be a volatilization source of a non-Aroclor congener, PCB11; average net flux of 14 ng m^{-2} day^{-1}.

480155 Efficiency of Bimetallic Nanoparticles for Treatment of Orange G in Water

Abstract:

Efficiency of Bimetallic Nanoparticles for Treatment of Azo Dyes in Water

Authors: Sheldon R. Shinn, Skylar Watson, Lauren F. Greenlee

The development of novel materials for water treatment applications remains a significant need as water resources become increasingly strained and contaminant removal challenges persist. While current commercial water treatment materials and technologies are able to remove a wide range of contaminants, the addition of functionality or performance through material design enables improved contaminant removal and the ability to address particularly recalcitrant contaminants. In particular, the development of nanostructured, reactive materials for water treatment applications holds great promise due to the reactive degradation and improved removal processes that have been demonstrated. Both laboratory-scale and pilot-scale research studies have demonstrated that iron-based nanostructured materials, such as iron nanoparticles, are able to remove a wide variety of water contaminants, including chlorinated organics, pharmaceutical compounds, and heavy metals, through both oxidative and reductive reaction mechanisms. However, iron nanoparticle lifetime remains a challenge particularly in oxidative environments, where the nanoparticles quickly react with water and oxygen to form non-reactive iron oxide compounds. Thus, there remains a need to investigate approaches to extend the lifetime as well as to enhance the performance of iron-based nanoparticles.

In our work, we have focused on the synthesis of bimetallic nanoparticles, which are then used to treat water contaminated by an azo dye (Orange G). These nanoparticles are synthesized in an aqueous solution and result in an alloy morphology. Variation of synthesis parameters can be used to optimize nanoparticle reactivity and understand which parameters are critical to controlling nanoparticle morphology. In our synthesis method, we focus on different combinations of metals, including iron-nickel and iron-cobalt. Thus far, our results have shown that different molar ratios of the metals result in different rates of removal of Orange G from solution, as well as in the overall removal of Orange G from the solution. Within a range of 0.1 to 0.4 mol Ni:mol Fe, lower molar ratios of nickel to iron in the nanoparticle metal combinations have shown higher removal of Orange G. In addition, changing the metals used in synthesis (e.g., from FeNi to FeCo) is a viable option to optimize the reactivity of iron-based nanoparticles. Finally, the types of stabilizers used during aqueous synthesis of the metal nanoparticles may impact the morphology and reactivity of the particles. Currently, stabilizers used in this work are poly(N-vinyl-2-pyrrolidone) (PVP) and amino tris (methylene phosphonic acid) (ATMP). However, other novel types of stabilizers, such as peptides and peptoids have proven to be successful in stabilizing nanoparticles in nanoparticle formation. However, it is unclear how the use of such specialized stabilizers impacts the reactivity of the nanoparticles in water treatment applications.

In this talk, we will present results on several synthesis parameters that have been identified as key to impacting nanoparticle reactivity towards the target water contaminant Orange G. Results will be presented for nickel to iron molar ratios in the range of 0 to 0.4 as well as iron to cobalt with a similar range of molar ratios. We will also present results on the characterization of these nanoparticles using techniques such as electron microscopy, x-ray diffraction, and x-ray photoelectron spectroscopy. Finally, results will be discussed as a function of nanoparticle concentration and Orange G concentration.

480323 Temporal Stability of Microbial Water Quality in Wye Island Irrigation Pond

Abstract:

Temporal Stability of Microbial Water Quality in Wye Island Irrigation Pond

A temporal stability study was conducted on an irrigation pond denominated Wye Island Pond in the state of Maryland with the goal of finding a relationship between bacteria and algae. Temporal stability is the study of points through time, in this case we studied where the E.coli concentrations are higher and lower than the average. The reason for studying this was to find points where bacteria was low, so consequently, irrigation from those points would reduce the amount of bacteria that gets to the crops. Additionally, irrigation water has recently been regulated under FDA FSMA and is a new interest in the food industry and technology. The objective of this study was to determine if a temporal stability pattern is present in the distribution of the microbial water quality indicator *E.coli*. In addition, algae concentrations were measured to identify a possible temporal stability pattern. Thirty-four different points were taken into consideration and were randomly studied three times on a period of two months. Twenty-four points were hand sampled from the banks and the other ten samples were kayak samples from the interior of the pond. After analyzing graphs, the bacteria and algae graphs demonstrated the same relationship throughout the pond. Meaning that the high concentrations were located approximately at the same area. But, when comparing both of them, relationship bacteria/algae, could not be found. Mean relative differences were computed and were found to be higher near the banks of the pond. This can be due to the water run-off depositing in the ponds and making the bank concentrations of bacteria higher, or can be caused by other natural phenomena. In the future there is a possibility of having more samples, by doing this, temporal stability concentrations could help us decrease the number of samples needed to test irrigation sites because of spatial trends.

479572 A Low Cost Solution for Recycling K-Cups and Reducing the Problem of Waste Plastic in American Landfills

Abstract:

A Low Cost Solution for Recycling K-Cups and Reducing the Problem of Waste Plastic in American Landfills

Shelby Browning and Jeffrey Seay, Ph.D., PE and Derek Englert, Ph.D., PE

University of Kentucky College of Engineering Paducah Extended Campus

As the amount of municipal solid waste (MSW) continues to increase in our environment, it is vital that new, more effective methods of converting this solid waste into useful substances be implemented. There are many industrial processes for transforming MSW, but there is no current process for transforming the plastic coffee pods known as K-Cups. K-Cups are made with a variety of plastic, one of which is known as number seven plastic. These K-Cups have a layered plastic design, making it near impossible for major recycling plants to use them. These two factors make it difficult to convert K-Cups effectively into useful substances. By utilizing a low-cost, easy to use processor developed by the University of Kentucky Appropriate Technology and Sustainability (UKATS) research team, K-Cups can be converted into an energy-dense fuel oil. Through this process, the amount of unwanted waste plastic, in the form of K-Cups, taking up space in landfills and harming the environment can be greatly diminished.

480542 Concentration and Microbial Dechlorination of Polychlorinated Biphenyls in Wastewater

Abstract:

Polychlorinated biphenyls (PCBs) are persistent organic pollutants that consist of a group of chemicals that can cause detrimental effects, when humans and animals are exposed to them in the environment. The chemical persistence and hydrophobicity of these molecules has rendered it difficult to quantitatively model their transport through the wastewater treatment process. The study site, the District of Columbiaâ€™s wastewater treatment facility (Blue Plains) is of interest since they must comply with the Environmental Protection Agencyâ€™s (EPA) PCB discharge limits.

The objective of this project was to analyze PCB monitoring data in order to develop methods of increasing PCB dechlorinating efficiency in the wastewater treatment process. Preliminary data shows that more sampling is necessary to avoid extrapolation errors. Additional process sampling will define methods for modeling PCB transformation in the wastewater treatment process.

480356 Improving the Estimates of Particulate Matter Impact on Human Health through Remote Sensing

Abstract:

Improving the Estimates of Particulate Matter Impact on Human Health through Remote Sensing

Wan Nurlaila Mat Desa[1], Gregory Carmichael[1,2]

[1] Department of Chemical and Biochemical Engineering, University of Iowa

[2] Centre for Global and Regional Environmental Research

Exposure to particulate matter severely affects human health as approximately 7 million people die from air pollution each year. There is an increasing need for accurate documentation of particulate matter in the atmosphere in order to provide more accurate estimates of exposure, and to document effectiveness of strategies to mitigate the adverse impact of pollution on human health. Currently surface based observations of particulate matter are done at only a limited number of locations within a city. Satellite based observations of aerosol optical depth (AOD) provide the possibility of adding more information on the spatial distribution of PM, which would increase the accuracy of exposure estimates. However the relationship between AOD and PM2.5 are complicated. This study examines how well aerosol behaviour can be predicted using atmospheric remote sensing data in Eastern Asia. To achieve the objective of the study, the relationship between PM2.5, PM10, and AOD was examined using observational data, taken from the KORUS-AQ campaign where samples from Gwangju station were selected for analysis. The relationships found in the observations were also compared to those predicted using the Weather Forecasting and Research (WRF) model.

Linear correlation between PM2.5 and AOD were determined from linear regression method. For each set of data obtained from Weather Forecasting and Research model (WRF-Chem) and observational (ground station and AERONET), plots of PM2.5:AOD to AOD values were constructed to give distribution slope (m) and correlation coefficient from the relationship of . The correlation of determination, R^2 obtained ranged from 0.167 to 0.530, while the slope ranged from -142.1 to -53.7. Linear regression function of observation PM to AERONET data showed the strongest correlation, with R^2 of 0.530, while model to model analysis only yielded 0.167. The direct comparison between the computed PM to AOD from yielded the lowest R^2 value. Strong correlation between observed PM and AERONET implies that measured AOD is a fair indicator of the PM concentration in Gwangju, compared to computed AOD using the WRF model. Values obtained from model simulations were higher compared to ground observations, indicating that more efforts are needed to improve the model capabilities. While this is the first step in improving the prediction of ambient particulate matter, the next step should incorporate space-based measurements that can provide broader spatial coverage.

480551 Eliminating a Flammability Risk in High-Level Radioactive Waste Processing

Abstract:

While working as an intern at the Savannah River National Laboratory, I completed laboratory-scale experiments in order to find processing conditions that minimized the production of hexamethyldisiloxane (HMDSO) during the processing of the high-level radioactive waste contained in forty-three 1 million gallon tanks at the Savannah River Site (SRS). HMDSO, a flammable gas, is a degradation product of antifoam 747, which is used to combat foaming during waste treatment at the Defense Waste Processing Facility (DWPF) at SRS. If unmitigated, the foam generated during processing could cause a foam-over and lead to contamination, a complete facility shutdown, and the loss of millions of dollars. The amount of HMDSO produced in the new Nitric-Glycolic processing flowsheet is greater than the amount seen in the current Nitric-Formic flowsheet. The Nitric-Glycolic flowsheet is planning on being implemented in DWPF in 2017, due to this flowsheet improving processing qualities such as reducing the amount of catalytic hydrogen produced. Being able to reduce the peak amount of HMDSO produced in the new Nitric-Glycolic flowsheet is important for the flammability strategy of the flowsheet. Working alongside chemists, statisticians, and engineers, the production rates of HMDSO in simulated, non-radioactive waste processing experiments were analyzed using a Fourier Transform Infrared Spectrometer (FTIR). The goal of these experiments was to reduce the maximum rate of HMDSO production to below the current safety basis. In order to both minimize flammability hazards and assist in HMDSO quantification, an air purge with a helium tracer was employed during experimentation. Process safety and interdisciplinary communication were key components in the success of the experiments. Four short-duration experiments were completed with varying operating conditions. In addition, a full-length, multi-day experiment was conducted in order to demonstrate that the revised antifoam strategy was effective throughout DWPF processing. The reduced antifoam strategy was deemed successful after demonstrating a significant reduction in HMDSO generation rates to within processing limits while simultaneously controlling foam. This antifoam strategy will be implemented alongside the new flammability safety basis for the Nitric-Glycolic flowsheet at DWPF in 2017.

480692 Biomass Fueled Multiple Effect Distillation Unit for Water Desalination

Abstract:

2016 AIChE Annual Meeting, San Francisco, CA, Nov 13, 2016

Biomass Fueled Multiple Effect Distillation Unit for Water Desalination

Ali Amiri, Kwabena Addae Sarpong, Michael Smith, O. John Idowu, Catherine E. Brewer

Department of Chemical & Materials Engineering, New Mexico State University

Water Science & Management Program, New Mexico State University

Department of Plant & Environmental Sciences, New Mexico State University

Abstract

Brackish well water is an issue that many New Mexico farmers face, and clean water for agricultural use can be expensive and difficult to obtain. The cost of clean water can affect the profitability of many crops and even influence what type of crops can be grown in certain areas. Thermal desalination is one method used to obtain clean water from available water sources. However, this method also requires a source of energy. Currently, desalination systems employ the use of sunlight, wind, and geothermal energy as potential energy sources. One resource that many farmers have available, but has not been thoroughly studied as an energy source for water desalination, is biomass. Through inert slow pyrolysis, biomass can be converted to bio-oil and biochar which can be used as fuel for a thermal

desalination unit and a soil amendment, respectively.

The purpose of this project is to determine the feasibility of a multiple effect distillation (MED) water desalination unit using biomass as an energy source. Results from this research will be useful for those with large amounts of unused biomass feedstock and are located in an area with brackish ground water.

480712 Re-Emission of Mercury from Activated Carbon in Coal Combustion Flue Gas

Abstract:

During the combustion of coal, mercury (Hg) is released and emitted from the stack into the atmosphere. After the release of Hg into the environment, it can react and form methyl mercury, which is the form of Hg that bio-accumulates in organisms such as fish and poses a significant health hazard to humans. Activated carbon (AC) has been shown to be a viable sorbent to capture Hg from coal combustion flue gas and is currently used commercially in power plants. However, untreated AC is not a very effective sorbent when compared to chemically modified/promoted AC treated with substances such as sulfur or halide salts. This study investigates Hg capture using a commercially available chemically modified activated carbon.

Currently there are concerns that AC has the capability of re-emitting already adsorbed Hg during unit start up due to ramps in temperature, sulfur trioxide (SO_3) concentration, or both. While the release of Hg is known to occur, the cause of re-emission is not well understood. The objective of this study is to investigate the factors that might have an impact on the re-emission of Hg from AC under a flue gas environment. In this study, the AC samples were exposed to Hg in flue gas composed of either only combustion products (N_2, O_2, CO_2 and H_2O) or a simulated flue gas (combustion products with NO and SO_2) to evaluate its capture capabilities. After being saturated with Hg in the flue gas mixture, the AC samples were subjected to a SO_3 concentration ramp (2 to 20ppm), a temperature ramp (from 105 °C to 205 °C) and both SO_3 and temperature ramp in order to understand how these conditions influence Hg re-emission. Temperature was found to be the main factor impacting re-emission resulting in the release of up to 40% of the Hg fed to the reactor. Ramping the SO_3 concentration independently had a negligible effect on emission/reemission, while ramping both the SO_3 concentration and temperature together showed a similar trend to that of the temperature ramp only.

480556 The Effect of Relative Humidity on Secondary Organic Aerosol Formation

Abstract:

Nano-scale particles in the atmosphere, known as aerosols, have important impacts on human health and in cloud formation. When aerosols are able to take on water under cloud-forming conditions, they are known as cloud condensation nuclei (CCN). Alpha pinene gas is emitted primarily by vegetation, and can react with a number of atmospheric components, including ozone gas, thereby generating secondary organic aerosols (SOA). The purpose of this investigation was to study the effects of relative humidity during the SOA-forming process on the CCN activity and particle size distribution of the products of the reaction of alpha pinene and ozone. This effect was studied by reacting atmospherically-relevant concentrations of ozone and alpha pinene in a one cubic meter Teflon smog chamber at varying levels of relative humidity and then sampling the resulting aerosol over time to measure the changing particle concentrations, size distributions, and ultimate CCN activity of the newly formed SOA.

480583 Synthesis of Magnetic Iron OXIDE CORE-Shell Nanoparticles with Tunable Coating Thickness for Enviromental Remediation

Abstract:

A method to synthesize magnetic iron oxide nanoparticles with a crosslinked polystyrene shell using surface initiated atom transfer radical polymerization (ATRP) is presented. The core-shell magnetic nanoparticles were developed to capture contaminents (e.g., polychlorinated biphenyls, PCBs) from contaminated water sources. First, Fe_3O_4 nanoparticles were prepared by reacting aqueous ferric chlorides with ammonium hydroxide. The initiator for ATRP, 2-bromo-2-methyl propionic acid (BMPA), was then coated onto the surface of the magnetic nanoparticles via ligand exchange. Finally, the BMPA-coated nanoparticles underwent ATRP process with styrene and curcumin multiacrylate (CMA) in organic media using copper and 4,4'-dinoyl-2,2'-dipyridil (DNDP) to carry out the controlled/"living" polymerization. The synthesized core-shell nanoparticles were then washed and stored in dimethysulfoxide (DMSO). Dynamic light scattering (DLS) was used to characterize the size of uncoated and coated nanoparticle as a function of reaction time. Fourier transform infrared (FTIR) spectroscopy, thermogravimetric analysis (TGA), and X-ray diffraction (XRD) were used to confirm the properties and presence of the polymeric shell. UV-visible spectroscopy was used to evaluate the stability of modified nanoparticles in DMSO solution. Overall, controlled shell thickness of the core-shell nanoparticles was demonstrated as a function of reaction time, obtaining nanoparticles with magnetic properties and shell network components that offer great promise for environmental remediation applications.

480730 Solar UV Disinfection of Drinking Water

Abstract:

According to the United Nations, nearly 800 million people (or 11% of the worldâ€™s population) lack access to an improved drinking water source. Water from unimproved sources can contain harmful pathogens, and over 2,000 children die every day due to lack of clean water, mainly in developing countries. An increasingly popular approach to solving this problem is to use solar ultraviolet light (which is abundant in many of the areas that lack clean water) to deactivate the pathogens. The purpose of this research project has been to better understand solar UV disinfection by developing effectiveness spectra for several key microorganisms. With the completed spectra, we know how much each wavelength of light contributes to disinfection, and are able to design more efficient disinfection processes. Samples of the strains *Salmonella typhi* LT2, *Vibrio harveyi*, and *Cryptosporidium parvum* were exposed to wavelengths of light ranging from 254 nm to 330 nm using a polychromatic, medium-pressure Lamp, and deactivation was tracked as a function of dose. It is well-documented that as wavelength increases above 254 nm, the disinfection becomes exponentially less effective, and the data from this project has maintained this trend. When the results were paired with data on how much UV radiation reaches earth from the sun, it was found that wavelengths between 310 and 320 nm are the largest contributors to disinfection for *S. typhi* LT2 and *V. harveyi*, whereas 330 nm contributes the most to deactivation of *C. parvum*.

480742 Analysis of Hydrothermal Carbon By Raman Spectroscopy

Abstract:

With the Flint, Michigan water crisis, lead contamination has become a visible concern in the public eye. The scale of the problem is gaining recognition, with over 2,000 water systems serving 6 million people in the US recently classified as contaminated with lead. While heavy metal remediation technologies such as coagulation and flocculation are feasible for developed countries, developing countries are limited in their ability to treat on a large scale.

Adsorption is a scalable, easy to implement technology for water treatment, with possible adsorbents such as silicates and activated carbon. Activated carbon is biomass or char subjected to high temperature gas treatment or acidic conditions, forming a denser carbon species with high surface area. Activated carbon, though versatile in its adsorption ability for both organics and metals, may be prohibitively expensive to implement on a large scale, and raises questions in the chemicals consumed and waste produced in its synthesis.

An alternative to this is hydrothermal carbon, or hydrochar, a carbon produced from treatment of biomass in low temperature water. As a result, hydrochar has a greater amount of acidic groups than AC, making it a promising candidate for adsorption of aqueous metals due to its great hydrophilicity and surface interactions. IR and Raman Spectroscopy were used to characterize hydrochar as a function of reaction time. This provides chemical information about the hydrochar that we will correlate to the adsorption behavior, allowing for us to design the hydrothermal process for production of better heavy metal adsorbents. To quantify our material as an adsorbent we have tested the copper adsorption capacity of our hydrothermal chars, and two industry grade activated carbons via UV VIS spectroscopy.

480755 Modeling Anaerobic Digestion of Biodegradable Solid Wastes

Abstract:

Anaerobic digestion (AD) is a series of biological processes that converts biodegradable organic matter into biogas. The US Air Force aims to use this process to treat the biodegradable solid waste produced on their forward operating bases (FOBs). The focus of this study was to obtain the Monod kinetic parameters of AD process intermediates (volatile fatty acids, monomers, and polymers) that are typically found in cardboard and food waste; the primary wastes produced on FOBs. These substrates were analyzed through initial rates experiments until 5% substrate was consumed to assume negligible growth. Granular sludge was used as the anaerobic inoculum. Monod kinetic parameters were obtained using experimental and computational methodologies. First, the AD of a single concentration of each of the substrates was regressed to obtain preliminary saturation constants (K_s) and maximum rates (μ_{max}). Next, these preliminary values were used to obtain actual experimental values. The experimental values obtained for the volatile fatty acids include: K_s = 0.237 g L^{-1}, μ_{max} = 0.000592 g COD g VSS^{-1} h^{-1} for acetic acid, K_s = 0.368 g L^{-1}, μ_{max} = 0.000475 g COD g VSS^{-1} h^{-1} for butyric acid, and K_s = 0.066 g L^{-1}, μ_{max} = 0.000155 g COD g VSS^{-1} h^{-1} for propionic acid. Likewise, the experimental values obtained for complex substrates were K_s= 0.0.92839 g L^{-1}, μ_{max} = 0.02679 g COD g VSS^{-1} h^{-1} for gelatin, K_s= 4.4157 g L^{-1} and μ_{max} = 0.00674 g COD g VSS^{-1} h^{-1} for cellulose. The parameter values obtained were within the range of previously reported literature values. Once all of the kinetic parameters of the most representative fractions of cardboard and food waste are obtained, they will be incorporated into the Anaerobic Digestion Model No.1 (ADM-1). ADM-1 predictions will be compared to actual experimental batch AD performance for food waste and cardboard, and model deficiencies will be discussed.

480767 A New Laboratory-Scale Device for Investigating Human Health Impacts of Biomass-Fired Domestic Cookstoves

Abstract:

Employment of simple cookstoves fueled by biomass continues to be a common practice for around 40% of the world population today1. Insufficient ventilation and use of unprocessed solid fuels contribute to indoor air pollution, which increases risk for diseases such as stroke, chronic obstructive pulmonary disease, and acute lower respiratory infection234. These account for over 4 million premature deaths per year.

Prior research on the health impacts of cookstove smoke inhalation mainly involve case studies, and only a few have focused on in-vitro/in-vivo experiments. The aim of this project is to design a device that enables stable production of smoke, mimicking a real-life domestic three stone cooking fire. This will allow for bench testing, that can be closely compared to the exposure one would see in a rural village that relies on these methods of cooking.

480773 Effect of Elevated Ammonia Concentrations on Secondary Organic Aerosol Formation during the Georgia Tech Ambient Field Study in Yorkville, Georgia

Abstract:

Secondary Organic Aerosols (SOA) are suspended particles formed in the atmosphere through the oxidation of emitted Volatile Organic Compounds (VOC). SOA make up a significant portion of ambient aerosol particles and therefore have an important impact on climate and human health. The formation of SOA is strongly influenced by inorganic species such as ammonia, however this influence is not well understood. Since ammonia emissions are expected to increase due to growing population and changing climate, we conducted an ambient field study to better understand the formation of organic aerosols in ammonia-rich environments. The study was conducted at a rural site in Yorkville, Georgia from August â€" October 2016. The Yorkville site is characterized by elevated concentrations of ammonia (10-20 ppb) due to a nearby cow pasture and high biogenic VOC emissions typical of the southeastern United States. The site is also part of the SouthEastern Aerosol Research and Characterization (SEARCH) network.

A suite of instruments was deployed to characterize gas and particle phase compositions over the period of the study. A Scanning Mobility Particle Sizer (SMPS) spectrometer consisting of a Differential Mobility Analyzer (DMA) and Condensation Particle Counter (CPC) was used to determine the concentration and particle size distribution of ambient aerosol particles in the size range of 10-300 nm. Additionally, a High-Resolution Time-of-Flight Aerosol Mass Spectrometer (HR-ToF-AMS) was used to determine the chemical composition of sub-micron non-refractory aerosol. Aerosol mass concentration was found to be relatively constant with slight diurnal variation. The particle size distribution for number concentration showed significant variation between day and night and revealed numerous nighttime growth events. The SMPS data were compared to AMS data to determine the chemical species present during episodes of high aerosol loading and growth events. Meteorological measurements from nearby SEARCH instruments were used to determine potential factors affecting mass concentration. From these and future results, the formation of SOA in the presence of elevated ammonia concentration will be evaluated.

477652 Implementation of the Argonaute Genome Editing Protein in Lactococcus Lactis

Abstract:

Probiotics are microorganisms that, when administered in adequate amounts, confer a health benefit on the host. For example, *Lactococcus lactis (L. lactis),* commonly found in dairy products, alleviates inflammatory bowel disease and has been engineered to excrete therapeutic proteins. However, the tools to genetically manipulate *L. lactis* remain limited and low-throughput. We hypothesize that the *Natronobacterium gregoryi* Argonaute protein (NgAgo) can be used as a genome editing method in *L. lactis.* Recently demonstrated to effectively promote genetic recombination in mammalian cells, NgAgo employs a complementary single-stranded DNA as a guide to cleave genomic DNA. By incorporating the NgAgo gene into *L. lactis*, we test the system to select for mutations at exact sights within the genome. Our work should result in a genome editing tool that is more selective and easier to use than the current industry standard, CRISPR/Cas9, with potential downstream applications in all-DNA-based rapid multiplexing technologies.

477653 Bacteriophage Activity on Paper Depends on Paper Properties and Immobilization Method

Abstract:

Bacteriophages are an excellent choice of substrate for detecting bacteria because of their specificity in binding to bacterial cells. Bacteriophages can be immobilized onto a surface which in turn can be used as a dip-stick test to detect concentration of bacterium. However, a problem with using phages for bacterial detection is finding an economical and producer friendly method for immobilizing the virions onto a surface. Some methods which have been utilized include direct addition, adsorption, and chemical binding of phage to a surface. Although, direct addition and adsorption have been successful in immobilizing phage there is still some question about the strength of adhesion of the viruses and whether or not they will remain adhered especially for long storage time or in harsh environmental conditions. On the other hand, chemical binding has proven to be a surer way to maintain immobilized phages on a surface. However, much of the chemical-binding research has been focused on utilizing smooth surfaces. The object of this study was to determine a method of chemically binding bacteriophage onto a porous surface, e.g. filter paper, and to determine the relative advantages and disadvantages of chemically binding bacteriophage versus merely adsorbing or directly adding bacteriophage to filter paper. We were able to utilize the 1-Ethyl-3-(3-dimethylaminopropyl) carbodiimide (EDC)/ *N*-Hydroxysuccinimide (NHS) cross-linker reaction to create amide bonds between the phage and the paper by coating the paper with chitosan and carboxymethyl cellulose (CMC) which provided the necessary carboxyl and primary amine groups needed. The results suggest that the use of polymer treatments and/or using the EDC/NHS cross-linker reaction for immobilization of the bacteriophage can increase the strength of immobilization of bacteriophage on a surface.

477690 A Look at the Recrystallization of Drug out of Supersaturated Amorphous Dispersions

Abstract:

70% of new drug candidates discovered in recent years show a lack of bioavailability due to poor water solubility.Â While new drug candidates are normally in the crystalline form due to that formâ€™s greater stability, the amorphous form has shown between 1.1 to 1000 times greater solubility. In order to stabilize the amorphous form and capture its solubility advantage, the drug can be suspended in a polymer to create an amorphous solid dispersion (ASD). While some factors have been identified that would lead to recrystallization of drug in an ASD, such as exposure to atmospheric H2O or sustained exposure to temperatures above the glass transition temperature, the behavior of a drug in an ASD remains a topic for research. This study looks at the behavior of a supersaturated ASD composed of Indomethacin (IMC) and PVP K12 as it is heated at temperatures where crystallization occurs. The data suggest that regardless of the initial amount of IMC in PVPK12, the IMC crashes out of the supersaturated solution to a consistent level based upon the amount of PVP K12 available to hold IMC. The data also suggest that as the annealing temperature increases the solubility of IMC in PVP K12 increases

Research Supported by the National Science Foundation REU Program #EED-1460486

477691 Encapsulation of UV Sensitive Substances in Polymer Particles and Related Coating Applications

Abstract:

ENCAPSULATION OF UV SENSITIVE SUBSTANCES IN POLYMER PARTICLES AND RELATED COATING APPLICATIONS

Protection from ultraviolet (UV) radiation is an important requirement for many products, such as pharmaceuticals, natural dyes, cars, and outdoor equipment. UV protective particles can be used to resist UV degradation by combining physical blocking and chemical absorption. Using spontaneous emulsion and homogenization methods, PMMA particles are synthesized containing the UV sensitive substance and protective materials (antioxidants). Various parameters are tested to optimize yield, entrapment efficiency, and particle homogeneity. Novel methods are being developed to coat the polymer nanoparticles on surfaces such as glass, cotton and polyester. UV/VIS spectrophotometry and microscopy image analysis reveal that a coating agent is required for UV protective nanoparticles to adhere to surfaces. Particle applications will continue to be tested using various coating agents, including PMMA and PLGA, as well as different UV sensitive substances to determine the effectiveness of UV protective particles and their potential use in industry.

477705 Kinetics of Growth and Biofilm Formation of Vibrio Cholerae in Presence of Potential PTS Modulating Chemical Compounds

Abstract:

Kinetics of Growth and Biofilm Formation of *Vibrio cholerae* in Presence of Potential PTS Modulating Chemical Compounds

Mahtab Waseem, Patrick Ymele-Leki

Department of Chemical Engineering, Howard University, Washington DC

Phosphoenolpyruvate-carbohydrate phosphotransferase system (PTS) is a multistep chemical process which regulates the intake and use of carbohydrates by bacteria. Bacteria growing through the use of nutrients, such as sugars, may form biofilms. Biofilms are multi-layered communities of bacterial cells that grow attached to one another. In addition to controlling sugar uptake by bacteria, the PTS regulates several cellular functions such as chemotaxis, glycogen metabolism, catabolite repression and the aforementioned biofilm formation.

Using a chemical screen, this study identified small molecule compounds that promote the production of *Vibrio cholerae* biofilms. *V. cholerae* MO10 served as model organism. It was grown in the absence of chemical compounds as a negative control, while an altered strain lacking PTS activity served as a positive control. Growth kinetics was evaluated in absorbance-based microplate assays, and quantitative biofilm assays were performed in borosilicate glass tubes. All assays were monitored by spectrophotometry. Biofilm induction was confirmed by monitoring the growth of *V. cholerae* in the presence or absence of chemical compounds. Further assays are underway to elucidate the mechanism through which candidate compounds affect the formation of biofilms. These assays will determine whether or not biofilm induction occurs through interactions with the bacterial PTS.

Success of this project will result in the identification and characterization of chemical agents that regulate biofilm formation by modulating activity of the bacterial PTS. This may lead to the development of novel microbial control strategies with applications in biological, medical, and pharmaceutical sciences.

Funder Acknowledgement(s): This study was supported by a grant from NSF (Research Initiation Award #1505301) awarded to Patrick Ymele-Leki PhD, Assistant Professor of Chemical Engineering, Howard University, Washington, DC 20059.

477834 Optimizing Targeted Drug Delivery to Polarized Macrophages Via Arginine-Based Surface Modifications of Liposomes

Abstract:

Tumor associated macrophages (TAMs) have been shown to facilitate tumor growth and share significant characteristics with macrophages alternatively activated by interleukin-4 (IL-4). Previous investigations of interactions between alternatively activated macrophages and doxorubicin loaded liposomes yielded promising results regarding the practicality of targeting TAMs for a customized immune response in combating certain types of adenocarcinomas. Liposomes, unilamellar lipid bilayer nanoparticles, have inspired extensive contemporary research for improving targeted drug delivery vehicles due to their biocompatibility, biodegradability, and tunable surface chemistry. In this study, two arginine-based compounds, 3-guanidinopropionic acid (B) and N-carbamyl-alpha-aminoisobutyric acid (I), were utilized to modify the surfaces of liposomes synthesized from 1, 2-dioleoyl-***sn***-glycero-3-phosphocholine (DOPC) and 1, 2-dioleoyl-***sn***-glycero-3-phosphoethanolamine (DOPE). Unmodified and modified liposome phospholipids were combined to synthesize 25%, 50%, and 75% modified liposome surfaces of each type and were subsequently loaded with doxorubicin, a well-documented cancer medication. The surface coverage of these modifications were thus varied in order to evaluate the minimum extent of surface modification needed for optimized macrophage internalization. Dilutions of these particles were incubated with three distinct RAW 264.7 macrophage phenotypes—naïve, classically activated via lipopolysaccharide (LPS), and alternatively activated via IL-4—to assess the half maximal inhibitory concentration (IC_{50}) of the liposome encapsulated doxorubicin.

477889 The Effects of Drug Treatment on Cellular Mechanics in Metastatic Breast Cancer Cells

Abstract:

Paclitaxel (Taxol), a common anti-cancer drug, has been on the market for over two decades, but its specific mechanism of action is not well understood. For this reason, it is not possible to calculate or personalize treatments beyond empirical knowledge of the maximum safe dosage. In order to work towards the possibility of personalized and calculated treatment, the specific effect of Taxol on cancer cellsâ€™ organization was examined and prepared for quantification. MDA-MB-231-Br cells were subjected to a spectrum of dosages of Taxol and an MTS assay to approximate the IC_{50} of Taxol on this cell line to be 10 nM. These cells were then treated with dosages ranging from 10 times above to 10 times below the IC_{50}, and stained for actin to visualize their actin cytoskeleton. Finally, a novel process was developed for building a detailed 3D finite element model of cancer cells from confocal fluorescent images. This project enables any work which currently uses a simplistic finite-element model to enhance results with a morphologically accurate model. In the future, the aforementioned process will be developed into a procedure for correlating and validating the model with experimental data, allowing predictive applications.

477816 Engineering Medium Chain Volatile Ester Biosynthesis By Eeb1 in S. Cerevisiae

Abstract:

Fatty acid ethyl esters are organic compounds produced during microbial fermentation and fruit ripening as secondary metabolites. These compounds have a wide variety of application in the food, cosmetic and biofuel industries. In particular, short and medium chain esters such as isoamyl acetate (a banana flavored fruit additive) and ethyl acetate (an industrial solvent) are naturally produced in the yeast saccharomyces cerevisiae (S. cerevisiae). The biosynthetic pathways that produces these esters involves a condensation reaction between an alcohol and an acyl- or acetyl-CoA by alcohol-O-acetyltransferases (AATase, E.C. 2.3.1.84 and 2.3.1.74). Previous high throughput screening of cell lysates revealed that Eeb1 exhibits high activity towards ethyl esters with acyl-CoAs of carbon chain lengths from C4-C12. However, the understanding of biosynthesis of medium fatty chain ethyl esters in yeast is not completely understood. Through overexpression studies, it has been revealed that Eeb1 localizes to the mitochondria, while mutations on the N terminus results in localization to lipid droplets. The mutations also increase the enzymatic activity of Eeb1 by 2.7-fold. This study presents the results of localization, studies activity assays, and western blot analysis to determine the relationship between localization, activity, and expression of Eeb1; which will help to further understand the biosynthesis of fatty acid ethyl esters in yeast. Future research will focus on the development of a biosynthetic pathway to efficiently produce octanoyl-CoA in cytosol and subsequently produce a high yield of octanoyl acetate under conditions of fermentation.

478088 Electrochemical Detection of DNA Methylation and Application to Breast Cancer Screening

Abstract:

Epigenetic changes, specifically an increase of DNA methylation at site specific regions, have been shown to correlate with around 80% of all cancers (Stratton et al., 2009). Traditionally, DNA methylation changes are detected using bisulphate treatment and methylation specific PCR. However, this technique is labor intensive, non-portable and does not facilitate efficient disease detection. The creation of a point-of-care electrochemical test would be both cost effective and clinically applicable. Using the eMethylsorb technique (Koo et al., 2014), disrupted covalent methyl modification of DNA was measured at a promoter region of a known breast cancer gene. The eMethylsorb technique takes advantage of nucleotides' differing affinities to gold. Post treatment unmethylated samples have a higher adenine composition compared to methylated samples. Adenine has the highest affinity to gold among nucleotides, and therefore, results in greater gold surface adsorption and a corresponding decrease in reactivity. The latter can be analyzed by measuring electrochemical redox reactions at the gold surface. To improve upon the eMethylsorb method, electrical impedance spectroscopy (EIS) was used to analyze surface changes at a stationary disc electrode and a rotating disc electrode (RDE). It was believed EIS would be a better method than cyclic voltammetry (CV) and differential pulse voltammetry (DPV) because EIS directly measures the charge transfer resistance whereas CV and DPV do so via an indirect manner. Previous publications have demonstrated EIS as an invaluable tool in the analysis of electrode surfaces and diffusive properties. EIS measures the signal response of a cell when an alternating current is applied at a specific frequency. The output, Nyquist plot, is plotted in terms of imaginary impedance. Stationary disc electrodes were found to produce unreliable results due to diffusive inconsistencies. Using the RDE, more reliable differences were found among the methylated and unmethylated samples. With a focus on controlling diffusion and maintaining a consistent electrode surface, the EIS method combined with RDE was found to be more repeatable when compared to previously published methods.

Koo, K.M., Ibn Sina, A.A., Carrascosa, L.G., Shiddiky, M.J. a., & Trau, M. (2014). eMethylsorb: rapid quantification of DNA methylation in cancer cells on screen-printed gold electrodes. Analyst 139, 6178–6184.

Stratton, M.R., Campbell, P.J., & Futreal, P.A. (2009). The cancer genome. Nature 458, 719–724.

478105 Engineering a High Throughput Method for Quantitative Analysis of Behavioral Neuromodulation in C. Elegans

Abstract:

For years, *Caenorhabditis elegans* has been a primary model organism for the study of biological processes, from development to aging. This species was the first multicellular organism to have its genome mapped and it is estimated that 60-80% of human genes have orthologs in *C. elegans*. The species has forty genes that code for different insulin-like-peptides (ILPs) which have been shown to influence various developmental and physiological processes, including lifespan and neural plasticity, but the specific roles of many of these peptides remain unknown. My research involves designing a high throughput platform to analyze the role that ILPs play in the behavior of *C. elegans*. This platform will allow for the quantitative analysis of movement, a downstream output of the nervous system, of a large number of organisms. The platform will be used to analyze the change in behavioral output of *C. elegans* lacking each one of the forty ILPs. RNA interference will be used to silence expression of the genes that code for the ILPs. This method involves feeding the worms bacteria that express double stranded RNA of the gene of interest and thus prevents the translation of the gene into protein. Videos of swimming animals will be acquired and analyzed by using specialized code in Matlab to quantify the differences in movement. This will be taking a step towards discovering what role each of the forty ILPs plays in modulating the state of the nervous system and learning about the ILP to ILP regulatory network.

477961 Optimizing an Electrospinning Process of Thermoresponsive Polymer-Protein Composite Nanofibers for Cell Culture

Abstract:

Electrospun nanofiber scaffolds have been used in tissue engineering applications because electrospinning is simple and the size of the fibers can be tuned based on set parameters in order to mimic the extracellular matrix. Poly(N-isopropylacrylamide), or PNIPAM, has proven thermoresponsive properties, which are attractive for cell culture; however, few studies have shown the ability of PNIPAM in 1,1,1,3,3,3-Hexafluoro-2-propanol (HFIP) solvent to be electrospun. Further, without crosslinking, PNIPAM is soluble in aqueous solutions, making it an inadequate choice as a biocompatible scaffold for in vitro research. Herein we discuss a method to electrospin, characterize and crosslink PNIPAM in HFIP. In order for the mats to be crosslinked, PNIPAM was co-spun with type I collagen in HFIP and then exposed to glutaraldehyde vapor. The crosslinked mats did not exhibit thermoresponsive behavior. However, cells survived and spread within the nanofibrous mats.

477995 Development of a Portable, Inexpensive Assay to Detect Antibiotic Resistance

Abstract:

While antibiotics have provided extraordinary health benefits, their overuse and misuse has consequently led to their diminishing efficacy. The U.S. National Library of Medicine and Center for Disease Control have declared antibiotic resistance a global health crisis, and have suggested improving diagnostic techniques and tracking methodologies in order to avert the crisis. However, taking these steps is difficult in developing countries where lab facilities are not readily available. We proposed the development of a portable, inexpensive, assay kit that uses solar energy to incubate and culture patient bacterial samples and, within 24 hours or less, deliver results regarding resistance to antibiotics. Bacterial growth can be measured in the form of a colorimetric assay that changes from translucent pink to bright yellow, which can be recorded using a smartphone camera and uploaded to a central database to track global antibiotic resistance. We have developed a 3-D printed portable device that can detect antibiotic resistance for a much lower cost than traditional molecular-based techniques. We have obtained data validating our colorimetric method using two strains of *Escherichia coli*, as well as *Staphylococcus epidermidis* and *Streptococcus pseudoporcinus*. Future work involves expanding the scope of this assay to detect resistance in other bacteria, such as *Staphylococcus aureus* and Group A *Streptococcus*, as well as developing the heating element to keep bacteria at optimum temperature for incubation in low-resource settings.

478023 Ligand-Induced Activation of an Engineered Viral Fusion Protein

Abstract:

Ligand-induced Activation of an Engineered Viral Fusion Protein

Introduction

Influenza is a pathogenic virus with various human and avian sub-types. In influenza, a membrane protein called hemagglutinin (HA) sits on the surface of the virus in an inactive state with its fusion peptide imbedded within the protein. The fusion peptide is a 20-amino acid sequence that plays a dynamic role in the infection of target cells. When exposed to a low pH environment, the protein will irreversibly change its structure so that the fusion peptide relocates and becomes exposed. HA is a model membrane fusogen and understanding its function will better inform attempts to design fusion activity into drug or gene delivery agents. The research objective is to characterize an engineered HA's activation in response to ligand binding and to determine if autocatalytic activation, previously observed for a HA of the H7 subtype, occurs in other HA subtypes.

Methods

Through Polymerase Chain Reaction (PCR), a primer encoded with a tetra-Cys motif replaced a section of the fusion peptide of HA and then was cloned back into the full sequence. The CCPGCC tag was placed at various starting locations within the fusion peptide's 20 amino acid sequence and cell lines of these mutants were created by transfection of Chinese Hamster Ovary (CHO) cells. By altering the sequence, it allows FlAsH, which is a fluorescent dye that binds to the tetra-Cys sequence, to bind to the fusion peptide when it is exposed. Once FlAsH binds, the dye fluoresces allowing the structural change and location of the protein to be seen within the cell via fluorescent video microscopy and to be quantified by flow cytometry.

Results

A 5-minute period of low pH exposure is necessary for the activation of a few HA on the cell surface, and, even after the removal of low pH environment, HA still activates fully on the surface of the cell. It was observed that without a low pH pulse, HAs were still being induced after a 7-minute incubation period with FlAsH, suggesting that FlAsH interacts with the fusion peptide upon binding to induce the conformation change. This was further tested by a real-time microscopy experiment of FlAsH and a neutral pH environment.

Discussion

The continuing activation of HA after the removal of an acidic stimulant suggests a disturbance in the membrane that causes all HA to activate fully across the cell or that some other type of communication occurs to create a cell-wide activation. The conformational change being driven by FlAsH provides an interesting avenue for activation of cell membrane fusion. It will make it possible to activate the conformational change without dependence on acid sensitivity but rather the presence of a ligand. These experiments were demonstrated with the avian influenza H7 subtype of HA. This work is currently being extended to the human-infectious influenza H3 subtype of HA. The dynamics of this subtype are proposed to be different due to the structure differences between the subtypes.

478226 Assembling Mucus-Penetrating Antibiotic Nanoparticles from Hydrophobic Ion-Pairing Based Flashnanoprecipitation of Hydrophilic Actives

Abstract:

There is currently no method to eradicate established pulmonary *Pseudomonas aeruginosa* infections among patients with cystic fibrosis. This in large part is due to the inability of antibiotics, such as polymyxin B, to penetrate thick lung mucosal layers formed by cystic fibrosis patients. FlashNanoPrecipitation (FNP) is a block-copolymer directed self-assembly technique that can encapsulate hydrophobic drugs into nanoparticles (NPs) that can carry drugs through mucosal layers. We here develop a new FNP process to encapsulate hydrophilic drugs using hydrophobic ion-pairing, where traditional FNP has been used to encapsulate strictly hydrophobic drugs. Sixteen ion-pairing reagents (IPs) were screened and used to successfully encapsulate several hydrophilic compounds and peptides, such as gentamicin and polymyxin. Ion-pair compositions can be varied to tune drug encapsulation efficiencies > 95%, and to tune drug release rates. Polymyxin B characteristic release rates vary from under five minutes to over seven days depending on the ion-pair:polymyxin encapsulation ratio, and ion-pair identity. The new mucus-penetrating FNP drug vehicles developed here creates new options for managing *Pseudomonas aeruginosa pulmonary* infections.

478264 Detection of Carotenoids Using Electrochemical Impedance Spectroscopy

Abstract:

Carotenoids are biological pigment molecules that are often found in fruits and vegetables. Humans cannot synthesize carotenoids, but they can be accumulated through diets (Fiedor & Burda, 2014). Carotenoids have been found to lower oxidative stress which can lead to a lower risk of disease and infections (Mares-Perlman et al., 2002). Lower carotenoid levels have also been found in patients with Alzheimer's Disease, Parkinson's Disease, and multiple types of cancer (de Munter et al., 2015; Nolan et al., 2014). Although there is a reverse correlation between the concentration of carotenoids and risk of disease, there are still a lot of qualities and mechanistic pathways of carotenoids that are unknown to scientists. An effective, point-of-care, cheap carotenoid sensor is needed to aid researchers in quickly and accurately quantify carotenoids in blood. A method for the detection of carotenoids was developed by using electrochemical impedance spectroscopy. Electrochemical impedance spectroscopy measures the impedance and frequency when an alternating current is applied to the system. A transfer of electrons was created by using the ferri/ferro cyanide reaction. Carotenoids, such as beta-carotene, were successfully detected by studying the interference of the carotenoid on the ferri/ferro cyanide surface reaction at a glassy carbon electrode. In order to increase the sensitivity of the detection of beta-carotene, the glassy carbon electrode was modified with cyclodextrin. Cyclodextrin has a pocket like structure with a hydrophobic interior and a hydrophilic exterior. Hydrophobic materials, such as beta-carotene, will be attracted and absorbed into the hydrophobic interior. Electrochemical impedance spectroscopy used with the modified glassy carbon electrode successfully detected small concentrations of beta-carotene.

Fiedor, J., & Burda, K. (2014). Potential Role of Carotenoids as Antioxidants in Human Health and Disease. Nutrients ***6***, 466–488.

Mares-Perlman, J.A., Millen, A.E., Ficek, T.L., & Hankinson, S.E. (2002). The body of evidence to support a

protective role for lutein and zeaxanthin in delaying chronic disease. Overview. J. Nutr. ***132***, 518S–524S.

de Munter, L., Maasland, D.H., van den Brandt, P.A., Kremer, B., & Schouten, L.J. (2015). Vitamin and carotenoid intake and risk of head-neck cancer subtypes in the Netherlands Cohort Study. Am. J. Clin. Nutr. ***102***, 420–432.

Nolan, J.M., Loskutova, E., Howard, A.N., Moran, R., Mulcahy, R., Stack, J., Bolger, M., Dennison, J., Akuffo, K.O., Owens, N., Thurnham, D.I., & Beatty, S. (2014). Macular Pigment, Visual Function, and Macular Disease among Subjects with Alzheimer's Disease: An Exploratory Study. J. Alzheimer's Dis. ***42***, 1191–1202.

478034 Gold Templated Indicator for Ionizing Radiation

Abstract:

This research addresses the need for improvement in radiation sensors/dosimeters during radiotherapy. The current sensors involved are polymer gel dosimeters, MOSFETs, and radio-chromic films. Most sensors involved require expensive equipment's and processing facilities for readout. There is still a need to develop better sensors that can be clinically applied. There are numerous groups around the world trying to conceive a better dosimeter. One of the radiation sensors that was developed recently was based on fluorescence signal emitted from the sensor. To advance the field of radiation sensors, we plan to use visual indication (colorimetric) as a method of detect ionizing radiation.

With an increase in cancer diagnosis in recent years, a treatment that is often prescribed is radiation therapy. When a patient undergoes radiation therapy, a portion of the patient's body is bombarded with ionizing radiation in hopes that the radiation will stop the growth of the tumor. A small difference in the radiation dosage can be a critical factor in the patient's battle against cancer. Unfortunately, even the smallest shift in the patient's body, such as breathing during the treatment process, can shift the tumor, and, therefore, reduce the amount of effective radiation dosage. This indicator would be injected into the tumor and extracted post-procedure to be analyzed. This will reveal the exact amount of incident radiation.

The chemicals that are used in the formula as of now are gold nanoparticles, $C_{16}TAB$, a surfactant, and nanopure water. The surfactant acts as a barrier between the gold nanoparticles and provides a method to keep them in the solution without thermodynamic precipitation. The Au^{3+} nanoparticles are diluted into the nanopure water and $C_{16}TAB$ in order to create a homogenous solution. Plasmonic nanoparticles, due to their unique physical and chemical properties, have received considerable attention in the recent past. Nanoparticles based on their formulation exhibit unique colors. This unique ability is the foundation behind our radiation sensors. Currently in literature there are numerous wet and dry methods to prepare nanoparticles one of which is using ionizing radiation.

Ionizing radiation splits water (solvent) into reducing and oxidizing agents. The reducing agents react with the metal ions and reduce them to zerovalent metals. The zero-valent metal, being unstable, aggregates to form larger nanoparticles. The formation of nanoparticles leads to an eventual color change that can be visually detected. In this case, we predict a linear correlation between change in color and the amount of radiation a sample receives.

A gold nanoparticle solution can be used to accurately determine the incident amount of ionizing radiation. We have developed a gold nanoparticle sensor in lab using Hexadecyltrimethylammonium bromide ($C_{16}TAB$) as the templating molecule. When this solution comes into contact with ionizing radiation there is a visible color change as the gold salt is reduced and templated, creating a dispersion within the fluid.

478269 Forward Genetic Screens for the Discovery of Functional Glucan Hydrolases in the Diatom Phaeodactylum Tricornutum

Abstract:

Phaeodactylum tricornutum is a marine diatom that is a promising candidate for advanced biofuel and bioproduct production. In order to further utilize this organism as a feedstock, it is crucial that researchers fully understand all aspects of the metabolic pathways involved in carbon storage and breakdown. Unlike many other species of algae, diatoms store sugars as chrysolaminarin, a polymer of glucose connected by Î^2-1,3 bonds with minimal Î^2-1,6-branching. Presumably, these polymers are hydrolyzed by unknown endo- and exo-1,3-Î^2-glucanases. To identify these putative hydrolases, this project employed a forward genetic screen with a cDNA library of *P. tricornutum* using *Escherichia coli* as a surrogate host. We first established that E. coli was not able to utilize laminarin, a practical equivalent of chrysolaminarin, as a sole carbon source. We then transformed *E. coli* with the cDNA library to pick clones that were able to grow on laminarin-medium. The clones presumably contain the candidate Î^2-glucanases of *P. tricornutum.* Preliminary findings include optimizing selective M9+Laminarin media, categorizing the growth behavior of *E. coli* on the media, and developing concrete screening parameters. This work may open doors for scientists interested in further engineering carbon fluxes in *P. tricornutum* for energy applications, in addition to researchers looking to further develop and expand forward genetics screens not only in this diatom, but other organisms as well.

478280 Measuring the Nanomechanical Properties of Tomato Cuticles Using Atomic Force Microscopy

Abstract:

Tomato plant (*Solanum lycopersicum)* consist of an outer skin surface called cutin. The cutin has two components: polymeric cutin and a mixture of waxes. Tomato fruit is subjected to stresses from environmental, biotic, and mechanical factors, leading to decreased crop yield and economic losses. It is important to understand the fundamental links between chemical composition and biomechanical properties of tomatoes. This project assesses the influence of cutin deposition on the nanomechanical properties of the tomato cuticle using Atomic Force Microscopy (AFM). The Youngâ€™s modulus of the glossy surface of dewaxed and waxy tomato outer cuticles from wild type (M82) and two RNAi-silenced cultivars (deficient in cutin synthase enzymes CUS1 and CUS2) were compared under controlled humidity (60%). The results were used to investigate if inhibition of the expression of the CUS1 and the CUS2 genes would result in significant differences in the elasticity of tomato cuticles. A downward trend in Young's modulus (M82, CUS2 and CUS1) was observed for the glossy surface of the dewaxed samples. This indicated that the elasticity of tomato cuticle is positively correlated to cutin deposition within the sample. Higher cutin deposition in the sample leads to higher values for the Youngâ€™s modulus. However for the waxy samples, CUS1 and CUS2 displayed a higher Young's modulus than M82. It was concluded that in the absence of epicuticualr waxes, as compared with the wild-type M82 cultivar, cutin deficiency in CUS1 and CUS2 contributes to a lower resistance to deformation and decreased stiffness of the cuticles in the absence of waxes. However, for the waxy samples the scenario is quite the opposite, with CUS2 and CUS1 having higher Youngâ€™s modulus values than M82. This trend can be attributed to the higher proportion of waxes to cutin in the CUS-silenced cuticles. In the future, these results will be correlated with findings from NMR and mass spectroscopy to better understand the underlying changes in chemical composition that produce cultivar-specific differences in nanomechanical performance.

478312 Comparison of MACS and FACS Sorting of Sca-1+ Cells from Mouse Bone Marrow

Abstract:

Cell based therapies offer a possible solution to some currently disastrous diseases. Some of the cells with a promising future for therapeutic use are biologically very rare, such as Hematopoietic Stem Cells (HSCs) and Endothelial Progenitor Cells (EPCs) which make up less than 1% of human bone marrow or peripheral blood. HSCs are marked with Stem cell antigen-1 (Sca-1) and are currently being used therapeutically to treat leukemia, lymphoma, and blood disorders. EPCs are able to regenerate blood vessels and are, therefore, hypothesized to be an effective treatment for many cardiovascular diseases. Effective and affordable isolation of these rare cell populations is necessary for new cell based therapies to be transitioned into clinical use. Of the two widely used antigen positive isolation techniques Magnetic Activated Cell Sorting (MACS) and Fluorescence Activated Cell Sorting (FACS), neither offers both high throughput and high purity sorting. However, a technique in development called Antigen Specific Lysis (ASL) is hypothesized to achieve these two criteria. This technique isolates cells by encapsulating the target population in a hydrogel, which also protects the target cell from lysis in hypotonic solutions. With the hope of comparing the three antigen positive cell sorting techniques for rare cell isolations, we first aimed to isolate Sca1+ cells from the bone marrow of mice using MACS and FACS. The cells isolated with these two techniques were compared for purity, recovery, and enrichment. We also compared the viability of the enriched cell samples and the amount of time it would take to sort a therapeutic dose of cells using each technique. Overall, the purity, recovery, and fold enrichment for FACS was higher than that of MACS. However, the time it would take to sort a therapeutic dose of EPCs from FACS is much longer than that for MACS. In our following study, we will be isolating Sca1+ cells by hydrogel encapsulation using ASL and compare the method against MACS and FACS.

Research Supported by the National Science Foundation REU Program #EED-1460486

478404 Process Development and Modeling of a Continuous Biphasic Liquid-Liquid System

Abstract:

Many pharmaceutical companies are beginning to adapt from batch to continuous manufacturing due to the numerous potential advantages it offers such as reduced waste, improved heat and mass transfer, and lower operational and capital costs. While continuous manufacturing has been widely used in the chemical industry, batch processes still dominate the pharmaceutical industry, presenting a series of regulatory, analytical, and process development challenges for adoption. Pharmaceuticals, as relatively low volume products, have generated a novel market for continuous microreactors, which still require sufficient understanding of scale-dependent parameters that can be illuminated via reactor characterization. Equipment characterization provides the ability to create enhanced process models, which are critical in process understanding and control to ensure an efficient, high quality product.

This particular study investigates a basic hydrolysis reaction in a continuous biphasic liquid-liquid system. In this system, the reaction kinetics are faster than the mass transfer rate, requiring an understanding of liquid-liquid mass transfer kinetics as a function of process variables. The reaction is performed in a Corning Low Flow (LF) reactor with extensive mixing and heat transfer performance as well as theoretical scale-up factors to the Corning G1 for pilot plant production. The objective of this project was to develop a transient model for the LF unit of the mass transfer coefficient (kLa) as a function of temperature, flow rate, and residence time for improved process control and understanding.

478038 Computational Fluid Dynamics Analysis of the Carotid Artery

Abstract:

In the human body, the carotid artery provides the brainâ€™s main blood supply carrying oxygenated blood from the heart to the head. As the carotid ascends the neck, it splits into two main branches referred to as the carotid bifurcation. This bifurcation contains a bulge, the carotid sinus, which causes a region of flow reversal to develop during the cardiac cycle. While this phenomena has been medically observed, it is not yet fully understood from a fluid dynamics perspective. The goal of this project was to characterize the velocity profiles and pressure gradients in the carotid artery to better understand the biological significance of the carotid sinus. Computational Fluid Dynamics (CFD - ANSYS Fluent) was used as the link between medicine and engineering, allowing us to quickly and easily visualize blood flow in this region under a variety of important physiological conditions. By creating a biologically accurate model of the carotid and applying relevant boundary conditions, blood flow and pressure contours were observed and applied to medically relevant issues. Overall, this research project served to help better characterize blood flow in the carotid artery. The simulations run in this experiment illustrated the same velocity profiles seen in previous studies, but in much more detail. The observed pressure contours also provided a cause for the flow reversal, showing an adverse pressure gradient within the carotid sinus. Baro- and chemo- receptors in the carotid sinus were also simulated to try and find biological significance for flow reversal. Further refinement is still needed, but these simulations show promise for future work. Applications of this research to atherosclerosis were then looked at and, as this is a medically relevant phenomena that is not fully characterized, this project could serve as a starting point for a much larger study. While this project helped to better characterize flow patterns and pressure gradients within the carotid, all simulations use average values and need verification with patient data. In the future, actual patient data could be used to model a blood flow in the carotid and serve to predict health problems. Future work will include working to obtain actual scans from patients and using their specific measurements to model actual arteries. The carotid bifurcation is very vulnerable to plaque build up, and as the carotid leads to the brain, these plaques breaking off are a major cause of blood clots and stroke. Overall, CFD allowed for a medical phenomenon to be visualized and better characterized with great promise

for future clinical significance.

LIQUEFACTION PROCESS OPTIMIZATION IN THE ENZYMATIC HYDROLYSIS OF STARCH MATERIALS

Acosta-Pavas, J. C[1]*., Betancur-Ramírez, K. J*[2]*., Bueno-Zabala, K. A*[3]*., Ruiz-Colorado, A*[4]

jucacostapa@unal.edu.co, kjbetancurr@unal.edu.co, kabuenoz@unal.edu.co, aaruiz@unal.du.co[1]

Grupo de Investigación de Bioprocesos y Flujos Reactivos, Departamento de Procesos y Energía, Facultad de Minas, Universidad Nacional de Colombia, Medellín.

Abstract

In this work, the optimization of conditions for liquefaction processes for two raw material are performed, starting with the characterization, measuring enzyme activity for SPEZYME®FRED of Genencor. Two central composite designs and response surface analysis were performed, for two variables, temperature and enzyme: substrate ratio, finally, statistical significance was determined from analysis ANOVA and R^2 of the design, and found the conditions that maximize the concentration of reducing sugars, temperature of 74.1°C and enzyme substrate ratio of 0.0276%, for pure wheat starch, for an expected value of 57.6 g/L with a confidence interval of 95 % between 52.51g/L-64.27g/L, for which validation was 61.89 g/L, and a temperature of 72.8°C and enzyme substrate ratio of 0.0516%, for Food Raw Material, for an expected value of 64,83g/L with a confidence interval of 95 % between 59.13 g / L-70.53g / L, for which validation was 60.50g /L, with the respective $R^2 = 94.16\%$ and $R^2 = 84.84\%$.

1. Introduction

An Industrial level, found process that generate subproducts or waste, these not being treated properly can cause various environmental problems, they are usually burned or dumped in landfills, releasing carbon dioxide, pollution of waterways and other negatives (Barragan et al., 2008).

Some of these residues are starchy, such as cassava, wheat, molasses, potato, banana, corn (Poonam & Ashok, 2011), which present potential food for obtaining glucose syrups. These residues are constituted by starch and this is formed by amylose, formed by glucoses linked by glycosidic bonds α-1,4 and amylopectin, consisting of glucoses linked by glycosidic bonds α-1,6 (Curvelo-Santana et al., 2010). Obtaining glucose is carried out by a three step process: gelatinization, liquefaction and saccharification. Gelatinization where grains are heated with water to generate the breaking of the starch and release amylose and amylopectin. Liquefaction carried out by amylases enzymes that hydrolyze the links α- (1-4) starch producing dextrins, maltose, maltotriose and maltopentoses. Saccharification, where glucoamylase enzyme (amyloglucosidase), hydrolyzes links α-(1-4) and α-(1-6) (Wild & Strain, 2011). Both enzymes are susceptible to inhibition by substrate, product, enzyme concentrations, temperature and pH. For this reason, the identification and optimization of reaction conditions affecting enzyme activity could improve the economic and technological viability of bioprocess (Wild & Strain, 2011). The objective on this work will make an optimization conditions for liquefaction processes for two raw materials, Pure Wheat Starch (Control) and Food Raw Material. Two central composite designs and response surface analysis were performed, for two variables, temperature and enzyme-substrate ratio (E:S Ratio). Finally, the conditions that maximize the concentration of reducing sugars were found.

2. Methodology

2.1. Characterization of Raw Material

For the characterization of two raw materials, we used Pure Wheat Starch and a mixture for two Food Raw Material, worked with a ratio 60:40 for sample 1 and sample 2; considering the starch content as the weighted mean of both samples.

We used the techniques for characterization of biomass, which were based on rules development by NREL, and the research group, Bioprocess and Reactive Flow, was performed, the techniques are:

1. Sample preparation (NREL/TP-510-42620)
2. Content of solids and humidity (NREL/TP-510-42621)
3. Ashes determination (NREL/TP-510-42622)
4. Extractive determination (NREL/TP-510-42619)
5. The amount of protein was determined by the Kjeldahl method of analysis (based on NTC 4657)
6. For the determination of Dry basis starch, used a Ewers Polarimetric Method (ISO-10520).

2.2. Enzymatic Activity Measure

The enzymatic activity measure of α-Amylase was based on adjusted method from enzymatic of α-Amylase (EC. 3.2.1.1) to SIGMA-ALDRICH; the procedure has been designed to measure the activity in Units for gram (U/g), it is defined that one unit releases 1.0mg of maltose to the starch in 15 minutes at pH of 6.9 and 95°C. The determination of starch hydrolysis products was done by the DNS method that measures the concentration of reducing sugars present in the sample.

From the Maltose curve was obtained a linear equation of the form $g_{maltose} = m * absorbance + b$, and the activity of α-Amylase is:

$$\frac{U}{mL_{enzyme}} = \frac{mg_{maltose} * fd}{mL_{enzyme}} \quad (1)$$

Where, the $mg_{maltose}$ obtained from the standard maltose equation by entering the absorbance of the test. fd is the dilution factor of the enzyme, and mL_{enzyme} is the volume of enzyme used in the reaction.

Finally, it became a specific activity, by calculating the density of the enzyme obtaining an activity as:

$$\frac{U}{g_{enzyme}} = \frac{mg_{maltose} * fd}{\rho_{enzyme} * mL_{enzyme}} \tag{2}$$

2.3. *Gelatinization Process*

A literature review was made in order to find the best conditions of temperature and time in the gelatinization process, and reported the table 1, these conditions were fixed; given that the main objective of this word was analyzed the liquefaction process. Based on the literature review and the data sheet of the α-Amylase enzyme, we selected, a temperature 95°C, and time 20 min, finally, according for pre-test, we change the time, 30 minutes for Pure wheat starch and 10 minutes for Food Raw Material, to guarantee the release of amylose and amylopectin.

Table 1. Conditions proposed for the gelatinization process

Conditions	**Gelatinization Process**
Temperature(°C)	66:110
Time(min)	5:15
Reference	Spezyme fred/optidex l 400; Nebesny et al., 1998 ; Ruiz et al., 2011; Betiku et al., 2013; Johnston & McAloon., 2014;Backs et al., 2008; Murthy et al., 2009

2.4. *Liquefaction Process*

A literature review was made in order to find the best conditions of temperature, time, pH, enzyme substrate ratio and reported the Table 2, based on the literature review and the data sheet of the α-Amylase enzyme, and pre-test performed, we selected, time in 60 minutes, solids percent en 20%, Calcium concentration 18ppm, pH 5.8, And a temperature range between 68:82°C and E:S Ratio between 0,02:0,011% for two Raw materials.

Table 2. Conditions proposed for the Liquefaction process

Conditions	**Liquefaction**
Temperature	40:110
Rpm	>400
Solids Percent (%)	NR
pH	5:7
Response variable	Reducing Sugar (g/L)

Time (min)	30:120
E:S Ratio (% p/p)	0.04:0.08
Calcium concentration	NR
Reference	Spezyme fred/optidex l 400; Ruiz et al., 2011; Betiku et al., 2013; Johnston & McAloon., 2014; Li et al., 2015

NR: Not Reported

2.5. Central Composite Design

2.5.1. Pure Wheat Starch

For Pure Wheat Starch, we performed a central composite design for two variables Temperature and E:S Ratio, the range were reported in Table 3.

Table 3. Range and levels of the independent variables for pure wheat starch

Variables	Range and levels		
	-1	0	-1
Temperature(°C)	70	75	80
E:S Ratio(%)	0.04	0.07	0.10

The conditions of the 13 tests performed are reported in Table 4, these tests were given by MINITAB® software trial version 15.0, program randomly.

Table 4. Order statistics in liquefaction process optimization for Pure Wheat Starch

Step	Temperature (°C)	E:S Ratio(%)	Step	Temperature (°C)	E:S Ratio(%)
1	75	0,07	8	75	0,07
2	70	0,1	9	82,07	0,07
3	75	0,07	10	75	0,07
4	80	0,1	11	80	0,04
5	70	0,04	12	75	0,07
6	75	0,028	13	75	0,11
7	67,93	0,07			

2.5.2. Food Raw Material

For Food Raw Material, we performed the Central Design composite for two variables Temperature and E:S Ratio, the range were reported in Table 5.

Table 5. Range and levels of the independent variables for Food Raw Material

Variables	Range and levels

	-1	0	1
Temperature(°C)	70	75	80
E:S Ratio (%)	0.04	0.07	0.10

The conditions of the 13 tests performed are reported in table 6, these tests were given by the MINITAB® software trial version 15.0, program randomly

Table 6. Order statistics for the liquefaction process of Food Raw Material

Step	**Temperature (°C)**	**E:S Ratio(%)**	**Step**	**Temperature (°C)**	**E:S Ratio(%)**
1	75	0,07	8	75	0,07
2	70	0,1	9	82,07	0,07
3	75	0,07	10	75	0,07
4	80	0,1	11	80	0,04
5	70	0,04	12	75	0,07
6	75	0,0299	13	75	0,111
7	67,93	0,07			

3. Results and discussions

3.1. *Characterization of Raw Materials*

The results of the characterization are presented in table 7, the content of Pure Wheat Starch 80.70% and the Food Raw Material 50.00%. With this, it is observed that a starch with a high content of starch (high purity) and a Food Raw Material with an medium content of starch. Starch is mentioned as it is the most important component in the liquefaction process and on which the optimization process will be carried out.

Table 7. Characterization of Pure Wheat Starch and Food Raw Material

Component	**Pure Wheat Starch**		**Food Raw Material**			
			Sample 1		Sample 2	
Humidity	7.87	0.22	2.34	0.11	2.71	0.03
Solids	92.13	0.22	97.66	0.11	97.29	0.03
Hexane Extractives	----	----	15.27	0.59	12.49	0.24
Water Extractives	----	----	12.58	0.46	12.70	0.60
Ethanol Extractives	----	----	6.38	0.12	4.76	0.04
Dry Basis Starch (Free Extractives)	**80.70***	----	**63.19**	----	**51.27**	----
Ashes(Free	----	----	0.83	0.05	0.68	0.10

extractives)						
Dry Basis Protein(Free Extractives)	2.30*	----	12.70	----	13.52	----
Total			110.94		95.43	

*With extractives

3.2. Enzymatic Activity Measure

The table 8 presents the results of the measurement of enzyme activity to α-amylase an enzymatic activity higher than reported by other authors is obtained, produced from the same microorganism. The unit of measurement that the commercial house performed on the enzyme, but which was not possible to compare with the unit commonly worked.

For the maltose curve, the concentration curve in g / L was performed, with respect to the absorbance, for which the following equation is generated: $y = 1{,}3751x + 0{,}0278$ with R^2 to 0.9998.

Table 8. Results enzymatic activity α –Amylase

Maltose (mg)	Dilution Factor	U/g enzyme	U/mL enzyme	Other Articles	Data Sheet
1	40411	2389	40411	18100U/mL**	17400Lu/g*

* Liquefon Units: One Liquefon Unit is the measure of the digestion time required to produce a color change with iodine solution, indicating a definite stage of dextrinization of starch substrate under specified conditions. ** Alfa amilasa de *Bacillus subtilis* from Genencor (Li et al., 2015).

3.3. Gelatinization and Liquefaction Process

3.3.1. Pure Wheat Starch

Table 9. Reducing Sugar Concentration in liquefaction process optimization for Pure Wheat Starch

Step	Reducing Sugar (g/L)	Step	Reducing Sugar (g/L)
1	**38,70**	8	**41,34**
2	35,31	9	29,37
3	**39,784**	10	**39,18**
4	30,56	11	44,94
5	46,21	12	**40,79**
6	61,85	13	32,28
7	38,63		

We found the concentration of reducing sugars, which will be taken as response variable. The central points have a minimum concentration of reducing sugars of 38.70 g / L and a maximum of 41.34 g / L.

3.3.2. Food Raw Material

Table 10. Reducing Sugar Concentration for the liquefaction Process of Food Raw Material

Step	Reducing Sugar (g/L)	Step	Reducing Sugar (g/L)
1	**59,37**	8	**62,15**
2	52,02	9	29,37
3	**59,60**	10	**61,11**
4	38,76	11	54,39
5	64,42	12	**67,60**
6	55,53	13	55,95
7	58,33		

We found the concentration of reducing sugars, which will be taken as response variable. The central points have a minimum concentration of reducing sugars of 59.37 g / L and a maximum of 67.60 g / L.

3.3. Central Composite Design

For the Central Composite Design of the Pure Wheat Starch and Food Raw Material, the respective ANOVA is reported in table 11.

Table 11. ANOVA liquefaction Process Optimization

Variables	Pure Wheat Starch p-value	Food Raw Material p-value
Constant	0.000*	0.000*
Temperature(°C)	0.037*	0.004*
E:S(%)	0.000*	0.112
T^2	0.013*	0.004*
$E{:}S^2$	0.014*	0.264
T*E/S	0.527	0.771

*Significant at < 0.05

According to ANOVA, the constant values and the temperature, the E: S Ratio and their square interactions, have a significant effect on the model for the case of Pure Wheat Starch, with respect to the Food Raw Material that presents a significant effect the constant, the temperature and square interaction. This doesn't mean that the E: S Ratio has no effect, only that for the selected range the enzyme can work with its best expected value.

The model equation for Pure Wheat Starch:

Reducing Sugar(g/L)= -54.9556+1.64280T-28.8817ES+0.0109912T2+299.732ES2-0.485039TES

With $R^2 = 94.2\%$ $R^2_{ajd} = 90.0\%$

The model equation for Food Raw Material:

*Reducing Sugar(g/L)= -1743.05+49.2195*T+671.977ES-0.336328T²-2734.95*ES²-5.37898*TES*

With $R^2 = 84.8\%$ $R^2_{ajd} = 74.0\%$

According to R^2 it is seen how the model fits better to Pure Wheat Starch than Food Raw Material.

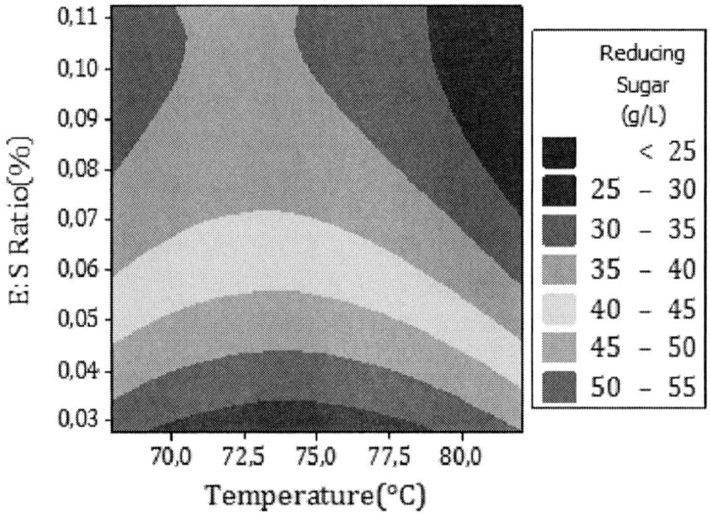

Figure 1. Surface Diagram, interaction Temperature and E:S Ratio with reducing sugars as the response variable for Pure Wheat Starch.

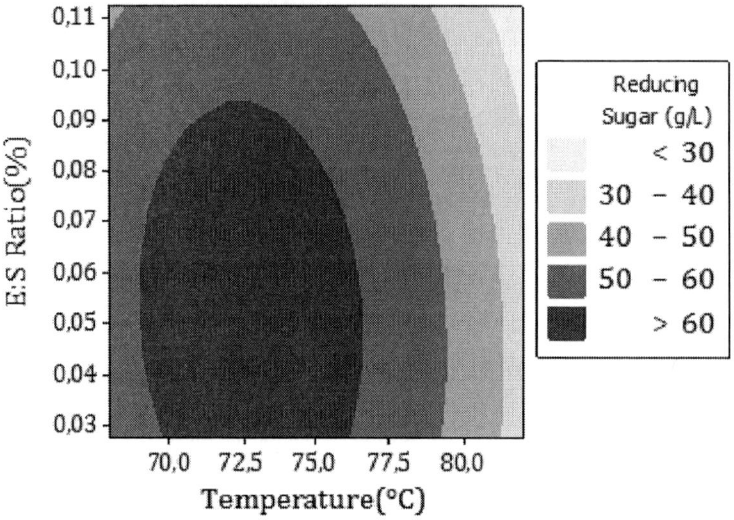

Figure 2. Surface Diagram, interaction Temperature and E:S Ratio with reducing sugars as the response variable for Food Raw Material.

For Pure Wheat Starch, Figure 1, presents as a temperature between approximately 70 ° C and 80 ° C, and an E: S Ratio between 0.04 and 0.10, it is observed how the temperature can generate a significant effect on the process and the E:S Ratio, for ranges lower than 0.03, can generate higher amounts of reducing sugars.

For Food Raw Material, Figure 2, the temperature has a significant effect on the process while for variations in the E: S Ratio, in that interval, there would not be a significant effect for variations in the concentration of reducing sugars, allowing to work as the minimum quantity with the aim of minimizing costs.

3.4. Validation Process

Table 12. Optimization process for Pure Wheat Starch and Food Raw Material

Optimun conditions Pure Wheat Starch Temperature=74.1°C E/S Ratio =0.0276%			
Reducing Sugar(g/L)	Confidence Interval	Expected Value	Validation
	52.51-64.27	57.6	61.89
Opmitum Conditions Food Raw Material Temperature=72.8°C E/S Ratio =0.0516%			
Reducing Sugar(g/L)	Confidence Interval	Expected Value	Validation
	59.13-70.53	64.83	60.50

It is observed that the Pure Wheat Starch has a lower expected value with respect to the Food Raw Material, due to the production process that has received this Food Raw Material, in relation to the high degree of purity of the pure wheat starch, which presents greater difficulty to obtain amylose and amylopectin from starch breakaging.

4. Conclusions

1. Optimal conditions for Control are temperature 74.1°C and E:S Ratio 0.0276%.
2. Expected value for control is 57.6g/L with a confidence interval of 95% between 52.5-64.3g/L. The validation was 61.9g/L.
3. Optimal conditions for Food Raw Material are temperature of 72.8°C and E:S Ratio 0.0516%.
4. Expected value for Food Raw Material is 64.8g/L with a confidence interval of 95% between 59.1-70.5g/L. The validation was 60.5 g/L.

5. References

Backs, T., Bruins, M. E., Matser, A. M., Janssen, A. E. M., & Boom, R. M. (2008). Effect of gelatinization and hydrolysis conditions on the selectivity of starch hydrolysis with ??-amylase from Bacillus licheniformis. Journal of Agricultural and Food Chemistry, 56(2), 488–495.

Barragán.H.B.E Tellez.D.Y.A Laguna. T. A. (2008). Use of agro-industrial waste. Laboratory Hazardous Waste. Department of Environmental Systems Engineering. National School of Biological Sciences. Av. Wilfrido Massieu S / N, Professional Unit Adolfo López Mateos. Mexico DF.

Betiku, E., Akindolani, O. O., & Ismaila, A. R. (2013). Enzymatic hydrolysis optimization of sweet potato (ipomoea batatas) peel using a statistical approach. Brazilian Journal of Chemical Engineering, 30(3), 467–476.

Curvelo Santana,J., Ehrhardt, D., Tambourgi, E. (2010). Optimizing of alcohol production from manioc starch. , 30(3), pp.613–617.

Ewa Nebesny, J. R. T. P. (1998). Enzymatic Hydrolysis of Wheat Starch into Glucose. Starch Stärke, 50(1979), 337–341.

Khawla, B. J., Sameh, M., Imen, G., Donyes, F., Dhouha, G., Raoudha, E. G., & Oumèma, N. E. (2014). Potato peel as feedstock for bioethanol production: A comparison of acidic and enzymatic hydrolysis. Industrial Crops and Products, 52, 144–149.

Li, Z., Liu, W., Gu, Z., Li, C., Hong, Y., & Cheng, L. (2015). The effect of starch concentration on the gelatinization and liquefaction of corn starch. Food Hydrocolloids, 48, 189–196.

Murthy, G. S., Sall, E. D., Metz, S. G., Foster, G., & Singh, V. (2009). Evaluation of a dry corn fractionation process for ethanol production with different hybrids. Industrial Crops and Products, 29(1), 67–72.

Poonam, S. N. Ashok P. (2011). Biotechnology for Agro-Industrial Residues Utilization. University of Ulser. Faculty of life & Health. United Kingdom.

Ruiz, M. I., Sanchez, C. I., Torrresa, R. G., & Molina, D. R. (2011). Enzymatic hydrolysis of cassava starch for production of bioethanol with a colombian wild yeast strain. Journal of the Brazilian Chemical Society, 22(12), 2337–2343.

Smrčková, P., Horský, J., Šárka, E., Koláček, J., Netopilík, M., Walterová, Z., … Hrušková, K. (2013). Hydrolysis of wheat B-starch and characterisation of acetylated maltodextrin. Carbohydrate Polymers, 98(1), 43–49.

SIGMA-ALDRICH. (2016). Enzymatic Assay of α-AMYLASE (EC 3.2.1.1). [cited 2016 May 5]. Available from: http://www.sigmaaldrich.com/

Spezyme fred/optidex l 400. Data sheet.

Wild, C. & Strain, Y. (2011). Enzymatic Hydrolysis of Cassava Starch for Production of Bioethanol with a Colombian Wild Yeast Strain. , 22(12), pp.2337–2343.

478485 Optimization of Process Variables for Design Ankyrin Repeat Proteins in Alzheimer's Diagnostic Applications

Abstract:

When normally healthy proteins in the brain begin to malfunction, they become misfolded and aggregate into conglomerates. These aggregations, called fibrils, are unable to be processed by the brain and therefore collect and cause damage to the surrounding neurons. As current forms of treatment are symptom based, there is little to no early-stage intervention. The diseases are tragically irreversible leaving little that doctors can do to help their patients. The development of a diagnostic system that can isolate the precursors to the degenerative effects will aid both researchers and patients to begin to better combat the disease.

Exploration of the applications of design Ankyrin repeat proteins (DARPin) has been used to combat Alzheimer's disease with protein engineering research. These particular proteins perform similarly to an antibody, and bind to a given antigen. DARPin are capable of being manipulated to have binding regions that target specific antigens. Previous studies have revealed that α-Synuclein protein aggregations within neural cells are related to these neurological diseases. Through application, the DARPin protein structures are capable of binding to these aggregation sites, producing diagnostic tests against these known antigens. The research presented herein focuses on optimizing processing conditions for production of α-Synuclein-targeting DARPin for use as a diagnostic in Alzheimer's patients. Once the presence of proteins is verified, the concentrations are measured using bicinchoninic acid assay (BCA). Enzyme-linked immunosorbent assay (ELISA) are used to analyze the functionality of the proteins. ELISAs are performed by stacking multiple layers of proteins and antigens, and then ultimately testing their presence and binding activity, using luminol. Various brain tissue samples (mouse, goat, and human) are used as targeted transgenic tissue for characterization. The brain tissue was genetically altered to overexpress α-synuclein to determine whether the synthesized proteins will bind well to their precursors.

The research has revealed that DARPin is optimized for a consistent growing temperature, regardless of protein structure. Utilizing a denaturing compound before running the proteins through SDS-PAGE gels improve the transfer through the media and aided in the characterization of the specific protein structure.

478518 Advancing Capstone Projects Beyond the First Generation: An Emergency Rapid Injection Device

Abstract:

Universities are a hub for innovation and design. Senior design courses often involve the development of a novel first generation prototype for biotech applications. However, due to the relatively short time frame, immaturity of the projects by the end of the year, and focus on curricular outcomes as precedence over end products, the courses rarely produce FDA-cleared or market ready products. This results in a plethora of novel ideas with significant promise, but no pathway to improve healthcare. The Bucknell Universityâ€"Geisinger Health System Medical Device Development Program (BUâ€"GHS MDDP) is a solution that allows a team of students to further develop products from senior design into second or third generation prototypes with a business plan for respective targeted markets. A multidisciplinary team of students in BUâ€"GHS MDDP was assembled, specifically based on technical skill sets needed to advance a previously patented but stagnant design project for an emergency rapid injection device, i.e. a syringe. This specialty syringe aims to deliver two sequential doses into an IV port with a single push. This injection is time sensitive because the drug, Adenosine, has approximately a 10 second half-life once entering the system. Preparation time associated with the current method of delivery is significantly longer, and therefore this new design provides time saving measures that could benefit the patient. Since the device was envisioned for use on Life Flight helicopters, students conducted a survey of medical staff members that regularly work on helicopters to better understand characteristics that would be valuable in this particular design. Initial feedback on the first design from professionals in the field showed that the majority felt there was a need for a new device and preferred the patented designs to the current method, demonstrating an interest and opportunity in the market for the device. A next generation functional prototype was created as a sequential step in the process and duplicated so that multiple medical staff could test it and give feedback on desired ergonomic and functional characteristics. The overall new prototype is closer to market, and connections have been made in the industry for possible future partnering.

478042 Analysis of Hypoxia in U87 Glioblastoma Suspension Cultures

Abstract:

Expanding cancer stem cell (CSC) populations while maintaining their vital stem cell character represents a significant challenge in modern biotechnology. CSCs are radio- and chemo-resistant, and further study requires a reproducible method for propagating them *in vitro*. Suspension culture holds promise as a paradigm for the expansion of cancer stem cells, since cells grown in non-adherent culture tend to maintain stemness more efficiently than those grown in adherent media. However, as cell spheres grow, their size can lead to hypoxic microenvironments in larger sphere aggregates, which can potentially preserve stem cell character, but can also be deleterious if the low oxygen levels lead to necrosis. Because of this, it is important for researchers to know the degree to which cells are exposed to hypoxia in culture. This work presents the implementation of novel methods for the characterization of hypoxia within multicellular tumor spheroids. A quantitative strategy was devised for determining the degree to which cells had been exposed to hypoxic conditions based on sphere volumes, as calculated by pre-processing images in ImageJ, then analyzing them using ReViSP, an open-source volume-estimation application for 2D images. Five samples of U87 glioblastoma cells were cultivated in suspension culture and incubated in ROS-IDâ„¢ Hypoxia Probe Reagent. The cells were then imaged and assayed for the presence of hypoxia using fluorescence microscopy, microplate spectrophotometry, and flow cytometry. A MATLAB-based script file was written and implemented to predict the degree to which cells had been exposed to hypoxia, based on the images collected. Sphere size distributions were built from the volume projections and used to predict hypoxia exposure for the cell populations, which was then compared to the experimental data.Â A correlation was found between the hypoxia detected by the fluorescence assays and the computational predictions. In the future, this strategy could be implemented in order to evaluate overall suspension culture hypoxia without performing labor-intensive assays.

478705 Gingivitis Prevention: Integration of Iron (II, III) Oxide Nanoparticles into Commercial Mouthwash

Abstract:

Gingivitis Prevention: Integration of Iron (II, III) Oxide Nanoparticles into Commercial Mouthwash

J. Nguyen, M. Osinski

University of New Mexico, Albuquerque, New Mexico 87120

2010 Data from the Center for Disease Control and Prevention estimated that 64.7 million of American adults are victims of periodontitis, a dangerous and irreversible periodontal disease. The early form of periodontitis is gingivitis, a reversible, yet widespread disease that is commonly overlooked. Therefore, experimentation is taking place to incorporate nanoparticles into dental care. Metal oxide nanoparticles have outstanding antimicrobial abilities, but may cause cytotoxicity within the human body. The high number of periodontitis patients and the biomedical limitations of nanoparticles have led to the need for non-toxic antimicrobial substances that are safe to be used in dental care products. A solution to this need may be found in the incorporation of iron (II, III) oxide nanoparticle in commercial mouthwash.

Iron (II, III) oxide (Fe_3O

eliminating bacterial biofilm. Using iron acetylacetonate ($Fe(acac)_3$) as our precursor, the $Fe(acac)_3$ was dissolved in triethylene glycol (TREG), which does not stimulate the immune system and made the nanoparticles water-soluble. After the nanoparticle synthesis, the behavior of these particle was then analyzed in water and Listerine Coolmint mouthwash.

The results showed that the nanoparticle synthesis was very efficient as it produced a 93% yield. Three characterization methods using dynamic light scattering, zeta potential, and transmission electron microscopy (TEM) were utilized. The analysis from the characterization methods confirmed that the average radius of one nanoparticle spanned approximately 5 nm, which is small enough to penetrate bacterial biofilm, and that all the particles displayed uniform size and shape distribution. The nanoparticles also displayed high colloidal stability in water. The same colloidal stability is predicted to be present within the nanoparticles dispersed in Listerine Coolmint mouthwash.

Further progress in our methods will require us to receive approval from the Institutional Review Board to obtain gingival swabs from gingivitis patients. The mixed culture bacteria from these swabs will further be propagated using agar on disks of hydroxyapatite (a prominent mineral found in human teeth). A series of matrix studies will be conducted with two control groups and an experimental group. All groups will be subject to the same environment on an orbital shaker in an incubator at human body temperature. The first control group will contain gingival bacteria subject to no treatment, the second control group will involve exposing gingival bacteria to plain Listerine Coolmint mouthwash, and the experimental group will include various concentrations of nanoparticles in mouthwash applied to gingival bacteria. Each group will be examined after increments of 10, 30, and 60 seconds. The viability of the gingival biofilm will be examined using the LIVE/DEAD BacLight Bacterial Viability Kit and cell culture size counting.

The characterization data from the Fe_3O

biomedically innocuous antimicrobial that, when combined with Listerine Coolmint Mouthwash, can increase the potency of the mouthwash. This creates a promising dental care treatment that can be easily incorporated into routine dental hygiene practices to effectively prevent the onset of gingivitis and periodontitis.

478538 Stability of Copper Modified Electrochemical Surfaces during L-Arginine Detection

Abstract:

L-arginine proliferative actions are a direct link to the onset of breast cancer in the presence of low nitric oxide levels (Singh et al., 2000). As a predictive biomarker for breast cancer, detecting L-arginine at a level of high sensitivity and accuracy is important. In previous research, metal nanoparticles have been used as catalysts, because of their ability to increase measurement sensitivity (Durst et al., 1997). With our electrode surfaces, we chose to modify it two ways with copper nanoparticles, either embedded into a polyaniline film (Cu-PANI) or in a hexacyanoferrate structure (Casella et al., 1996; Chen et al., 2007). We first tested the sensitivity of the Cu-PANI on a glassy carbon electrode, providing preliminary data indicating the removal of copper from the PANI surface. Through the use of an electrochemical quartz crystal microbalance (eQCM), I was able to validate our preliminary observation. The eQCM is helpful in measuring mass changes by monitoring the quartz frequency changes that occur on the quartz surface. Our concerns with the Cu-PANI stability prompted further examination to understand the reasons for the delamination. In the end, we were able to confirm four factors that increased the speed of the delamination process including pH, flow rate, voltage, and concentration. Ultimately, the factors had little affect on the copper hexacyanoferrate and, in comparison to Cu-PANI, had less copper removal. Our focus on the modification's stability will continue for the purposes of detection of L-arginine for later comparison of stability versus the sensitivity.

Casella, I.G., Cataldi, T.R.I., Guerrieri, A., and Desimoni, E. (1996). Copper dispersed into polyaniline films as an amperometric sensor in alkaline solutions of amino acids and polyhydric compounds. Anal. Chim. Acta ***335***, 217–225.

Chen, S.-M., Li, S.-H., and Thiagarajan, S. (2007). Preparation of Bilayer Platinum and Copper Hexacyanoferrate Hybrid Film Modified Electrode and Its Electrocatalytic Properties. J. Electrochem. Soc.

154, E123.

Durst, R.A., Baumner, A.J., Murray, R.W., Buck, R.P., and Andrieux, C.P. (1997). Chemically modified electrodes: Recommended terminology and definitions. Pure Appl. Chem. ***69***, 1317–1323.

Singh, R., Pervin, S., Karimi, A., Cederbaum, S., and Chaudhuri, G. (2000). Arginase activity in human breast cancer cell lines: N(omega)-hydroxy-L-arginine selectively inhibits cell proliferation and induces apoptosis in MDA-MB-468 cells. Cancer Res. ***60***, 3305–3312.

478707 Delivery of Model Lipophilic Compounds from Poloxamer 407 Thermoreversible Gels

Abstract:

Permeation into the skin from topical drug application favors highly lipophilic drugs due to the low water content and intercellular lipids of the stratum corneum (outermost layer of the skin). However, many lipophilic drugs are BCS Class II or IV with low water solubility that limits their bioavailability in the hydrophilic dermal environment. Thermoreversible poloxamer gels are well tolerated topical delivery systems that can concentrate drug in the epidermis and dermis. Addition of excipients and drugs can significantly affect the solution to gel transition temperature ($T_{sol \rightarrow gel}$), and the effect may be dependent upon the physicochemical properties of the active ingredient. In this work, two model lipophilic compounds, loperamide hydrochloride and diclofenac sodium, were used to formulate thermoreversible gels. These compounds were chosen for their comparable lipophilicity (logP 4.51 and 5.5 for diclofenac and loperamide, respectively) but varying water solubility (0.0086mg/mL and 50mg/mL for loperamide and diclofenac, respectively). Gels were prepared with a poloxamer 407 base to make thermoreversible aqueous solutions. Due to its nearly negligible water solubility, the mixed solvency concept was initially used to enhance the solubility of loperamide in aqueous solutions. Maximum loperamide solubility occurred in a 40% solution of propylene glycol (10 w/w %) and PEG 400 (30 w/w %), increasing its solubility by nearly 1,500 times compared to water. Solubility results informed formulations of thermoreversible gels, prepared using the cold technique with poloxamer 407. A loperamide-incorporated gel (16% w/w Poloxamer 407, 0.5% loperamide, 8% PEG 400, 24% propylene glycol, 4% ethanol) and two diclofenac sodium-incorporated gels (17% poloxamer 407, 0.5% or 1% diclofenac sodium) were investigated for solution to gel transition temperatures ($T_{sol \rightarrow gel}$) and *in vitro* drug release. Solution to gel transition temperatures were 23 ± 3.5°C, 29.3 ± 0.6°C, and 31°C for the 0.5% and 1% diclofenac and 0.5% loperamide gels, respectively. *In vitro* drug release studies were conducted using excised porcine skin and resulted in an unexpectedly rapid lag time of less than three hours for both drugs.

478551 Obtaining and Purification of Crude Extracts with Cellulase and Laccase Enzymatic Activity from an Isolated Strain of Fusarium Spp

Abstract:

The growing demand of biotechnological applications requires the use of microorganisms of interest, because of its production of molecules with applicability in industrial processes, or its favorable interaction with other microorganisms; in order to maintain living organisms for a certain time, and preserving their own morphological characteristics (Voyron et al., 2009) protocols should be implemented to generate low variations in phenotypic and genotypic aspects, with less impact on its added value. The genus ***Fusarium*** is a type of filamentous fungus which is distributed in a wide number of ecosystems. Due to its characteristics, it is part of the soil microbiota and can also be found in association with plants and other organisms. Filamentous fungi supplement their nutritional needs from biomass degradation, which makes them organisms of interest for its production of enzymes for cell wall degradation. Cellulose is one of the most important renewable sources of carbon, which is mostly synthesized by plants, with an annual production of 7.2×10^{11}tons (David Krogh, Rock, Åkesson, & Nielsen, 2005).

In this paper optimization for ***fusarium spp.***specific medium was performed. In order to favor the Laccase production, methods of inoculation with PDA solid medium, then in PDB liquid medium, and finally a modified Kirk culture medium for 8 days and 200 rpm at 30 ° C, were applied. This extract was vacuum filtered with a pore size of 0.2 micron, and finally stored at low temperatures for measuring its enzymatic activity for Laccase, in international units, and Cellulase in F.P.U units. The strain preservation was carried out by immersion in sterile mineral oil at room temperature.

After production of the extract, the process of separation and purification of the enzyme was performed using three-phase partition method. For this experiment, was made central composite design, which had input factors of temperature, tert-butyl alcohol (extract: tert-butanol) concentration and ammonium sulfate saturation (% PV), depositing the phase of interest in a phosphate buffer, maintaining its pH at 7 for one hour. Finally, the extract was measured in FPU Cellulase activity, using the DNS method.

The extract was produced with Xylose as a carbon source, peptone as nitrogen source and methanol as inductor, presenting as higher enzymatic activity 42 U/L. By purifying the Laccase from the produced extract, a pure enzyme with fold purification 1.2164 ± 0.1830 of performance and 114.13 ± 7.69% was obtained. For this same extract, Cellulase activity was tested, obtaining a 0,08 U/L; this result was verified for specific activity in FPU, showing that Cellulase activity was not significant, hence a growth medium for this strain was suggested in order to increase the Cellulase activity in the extract.

References

David, H., Krogh, A. M., Roca, C., Åkesson, M., & Nielsen, J. (2005). CreA influences the metabolic fluxes of Aspergillus nidulas during growth on glucose and xylose. ***Microbiology***, ***151***(7), 2209–2221. http://doi.org/10.1099/mic.0.27787-0

Van Den Brink, J., & De Vries, R. P. (2011). Fungal enzyme sets for plant polysaccharide degradation. ***Applied Microbiology and Biotechnology***, ***91***(6), 1477–1492. http://doi.org/10.1007/s00253-011-3473-2

Voyron, S., Roussel, S., Munaut, F., Varese, G. C., Ginepro, M., Declerck, S., & Filipello Marchisio, V. (2009). Vitality and genetic fidelity of white-rot fungi mycelia following different methods of preservation. ***Mycological Research***, ***113***(10), 1027–1038. http://doi.org/10.1016/j.mycres.2009.06.006

Rodríquez-Siza, Jose Alejandro. (2015). "Estudio Previos En El Pretratamiento Enzimatico Con Lacasa Para Cascaras de Cacao." Universidad Nacional de Colombia sede Medellín.

Certain, L. (2015). Optimización del medio de cultivo para la producción de lacasa por ***Fusarium*** sp. Universidad Nacional de Colombia sede Medellín.

478559 Photoacoustic Imaging of cRGD Targeted Nanoparticles in a Lewis Lung Carcinoma Tumor Model

Abstract:

Photoacoustic (PA) imaging is a biomedical imaging modality that relies on photon adsorption and processing of resultant pressure waves by conventional ultrasound transducers to construct images in 3D real-time high spatial resolution. Through PA imaging, multiple species can be simultaneously imaged; however, these efforts have been greatly limited by the availability of spectrally separable contrast agents that can be used in vivo. Towards this end, we present here the development of a mixed nanoparticle (NP) system for real-time simultaneous in vivo multiplexed PA imaging. Nineteen water-insoluble organic dyes were screened for poly-ethylene glycol (PEG) coated nanoparticle stability, yielding ten with unique separable absorbance profiles, highly tunable optical properties within the near-infrared window, and in vivo compatibility. This new in vivo compatible mixed-NP system with unique separable absorbance profiles enables advanced PA imaging applications with pharmacokinetic implications. Mixed-sample cRGD peptide surface modified NPs and non-modified NPs were simultaneously tracked in real-time in a murine subcutaneous Lewis lung carcinoma tumor model. Simultaneous imaging of non-modified NPs serves to normalize the effects of active-ligand against passive-NP targeting effects with, for example, 25% surface cRGD modified NPs displaying only 9.8 ± 4 fold higher liver to tumor ratio accumulation levels as opposed to 52 ± 22 fold accumulation levels pre-normalization, questioning the effectiveness of NP targeting. As a result, our new nanoparticle contrast agent tools advance the field of PA imaging with in vivo real-time multiplexed imaging capabilities and pharmacokinetic applications.

478560 High-Efficiency Capillary-Driven Plasma Separation with Porous Filter Materials for Point-of-Care Plasma Sampling

Abstract:

This work presents the first capillary driven microfluidic device that separates a specific volume of plasma from a blood sample of unknown volume. We demonstrate high plasma extraction efficiency from whole blood using capillary forces, where approximately 20μL of plasma can be extracted from 50μL of whole blood.

Blood plasma-based tests are the gold standard for analytical laboratory analysis, owing to their ability to derive a large range of information pertaining to patient health. However, contemporary tests require collection of venous whole blood samples in milliliter volumes, where they are fractionated via centrifugation to obtain plasma. The disadvantages associated with this method include invasiveness, costliness, collection of large volumes of blood, slow processing times, and inability to be performed in varied settings. To alleviate these issues, an autonomous microfluidic device was developed to extract blood plasma. This functionality was achieved by combining a porous filtration membrane with a hydrophilic microchannel. Using such a device, various commercially available filtration membranes were characterized for their extraction capability, efficiency, and extraction kinetics.

Of the 7 commercial membrane filters tested, 4 were successful in extracting plasma from untreated whole blood while 3 filters demonstrated massive leakage of blood cells and hemoglobin. Filtration rates and filtration efficiencies were measured to determine the most effective filtration membrane. We found that the Vivid Plasma Separation membrane, model GR was most suitable, yielding 75% plasma extraction efficiency within 10 minutes. Additionally, we found that membrane filters with coatings, used to minimize cell lysis by osmotic effects, yielded higher plasma extraction volumes compared to the uncoated filter; however the increased volume may be due to dilution of plasma. This potential source for bias has to be considered when capillary extracted plasma is used for quantitative downstream analysis. Future work to be completed is to fit the extracted plasma volume to the channel dimensions and dissolvable

valve characteristics, as well as to optimize the timing of deviceâ€™s functional sequence.

478566 Dielectrophoretic Response of Condensed DNA Clusters in AC Fields

Abstract:

The use of insulator-based dielectrophoresis (iDEP) for size-dependent separation and fractionation of DNA fragments has been regularly applied to lap-on-a-chip devices. During the use of insulator-based dielectrophoresis, homogeneously dispersed DNA molecules have been known to reversibly associate into segregated clusters under the influence of strong AC electric fields at low frequencies. This recently observed clustering phenomena has since gained attention in the field of colloidal physics, however, the significance of this visible DNA clustering behavior for DNA dielectrophoresis has yet to be investigated. Under these conditions, clustering behavior is believed to influence dielectrophoretic sorting for DNA molecules. As a result, this phenomenon could be used to improve DNA sample preparation and pre-concentration using microdevices for Next-Generation Sequencing (NGS) applications. Research investigating cluster formation through modeling and computer simulations has shown that this phenomenon may be due to electrohydrodynamic instability resulting from dipole-dipole interactions between molecules. Clustering behavior has the potential to enhance the efficiency of DNA fractionation by exploiting strong AC electric fields to induce condensation and separation of DNA molecules. Here, we examined DNA cluster formation and migration characteristics using various electric field parameters. Various amplitudes and frequencies of applied potential were examined to determine the extent of correlation in DNA clustering. DNA clustering was largely observed with frequencies ranging from 10 to 100 Hz and electric field strengths above 800 V/cm. Similar electric field parameters have produced some of the highest-observed sorting efficiencies in an iDEP constriction sorter for DNA molecules ranging in size from 10 - 50 kbp. These clusters correspond to unique dielectrophoretic behaviors for large DNA fragments. Our work suggests that DNA clustering can be exploited using insulator-based dielectrophoresis for size-based separation and fractionation with extremely high efficiencies. Such high efficiencies will significantly advance sample preparation and concentration of DNA and other macromolecules for biosensing and bioanalysis.

478571 Expression and Purification of Plant Embryogenic Proteins

Abstract:

The main goal of this project was to develop and optimize an efficient, scalable procedure for large-scale production and purification of Phytocyanin-Like Arabinogalactan (PLA) proteins expressed in *Escherichia coli* from various plant species. PLA proteins are a family of proteins that have shown improvement of somatic embryogenesis and regeneration in *Gossypium hirsutum* (upland cotton) (*Poon et al*). Poonâ€™s work was furthered to explore the somatic embryogenic potential of *Theobroma Cacao* PLA (TcPLA), an ortholog of GhPLA. Accomplishing industrial-scale production of high quality PLA proteins would benefit crop improvement programs â€“ ultimately improving nutrition and yields in the face of increased population and degrading environmental conditions. This research project, funded by the National Science Foundation and the Bill and Melinda Gates Foundation, explored an alternative purification method of PLA proteins to optimize the production process.

478573 The Analysis of Keratins/Chitosan Photocrosslinked Biomaterials for Bone Tissue Repairing

Abstract:

Bone tissue engineering has gained increasing attention among the medical society recently due to the aging population in developed countries. Issues of bone repair clinically is very popular, especially regeneration of large traumatic bone defects. The materials used in bone tissue engineering were dominant by either inorganic ceramic compounds or synthesis polymers. But natural polymers are also of great interest of researchers in this biomaterial field since they posses intrinsic biocompatibility, biodegradability and with little immune response. These types of natural polymers, for example, collagen, keratin, alginate and chitosan, are being extensively studied for their potentials in tissue engineering recently.

Keratin has intrinsic biocompatibility and contains several peptide-binding motifs that support the attachment of a wide variety of cell types. Based on our previous study, we have proved that keratin extracted from human hair can promote cell adhesion and proliferation for 3T3 fibroblasts, MG63 osteoblasts and human adipose stem cells (hASCs). Despite the advantages of keratin in bioactivity, the fragile mechanical properties of keratin will face challenges in the field of bone tissue engineering. To overcome this problem, we decided to combined keratin with chitosan by induction of azide functional group as a photocrosslinker, in order to enhance the mechanical strength.

In this research we will analyze the properties of the Keratin/Chitosan composites, and prove that the following films can maintained good adhesion and proliferation of human adipose stem cells. Most importantly, this biomaterial could promote the differentiation to osteoblast. Also, we will show that by lyophilizing the keratin chitosan mixtures, a scaffold of highly interconnected porous structure can be made, which may produce an ideal environment for tissue engineering. .

We believe that the keratin/chitosan composite biomaterials can be used in surface modification, and the crosslinkable properties can make scaffold fabrication of these natural polymers for bone tissue engineering possible in the near future.

478577 Structural Investigations of Caffeine Metabolic Enzymes

Abstract:

Pseudomonas putida CBB5 digests caffeine and related purine alkaloids through an *N*-demethylation pathway. Five key enzymes have been found responsible for this: NdmA, NdmB, NdmC, NdmD, and NdmE. Classical bioengineering techniques were used to further investigate these enzymes. When a His_6-tagged NdmB and a non-His_6 tagged NdmA were co-expressed in *E. coli* and cell lysate passed through a Ni-NTA column, both were retained on the column and co eluted with 500 mM imidazole. These results indicate that NdmA and NdmB form an enzymatic complex when coexpressed *in vivo*. The ultimate goal of this project is to create chimeric enzymes combining NdmA or NdmB with NdmC. Toward this end, homology models of the enzymes were analyzed and the chimeric enzymes were designed *in silico*. The chimeric genes have also been cloned into *E. coli* expression vectors and activity assays are pending. Caffeine and its metabolites are in high demand due to their use in food, beverages, and pharmaceuticals. By understanding more about the enzymes themselves, it is expected that the reaction pathways can be better controlled for more efficient yields of valuable biochemicals.

478579 Uranium Concentration Analysis By Visible Absorption Spectroscopy- Temperature and Solution Matrix Control

Abstract:

Mo-99 (66 h half-life) decays to Tc-99m, the dominant medical isotope used in nuclear medicine. The Office of Material Management and Minimization (M3) within the National Nuclear Security Administration (NNSA) has an objective to 'accelerate the establishment of a reliable, commercial Mo-99 supply network that avoids a single point of failure and does not use proliferation-sensitive HEU (High Enriched Uranium)'. This remit includes a cooperative agreement with SHINE medical technologies, a company aiming to produce Mo-99 domestically in the US. SHINE plans to use a deuterium-tritium neutron generator to produce fission Mo-99 from Low Enriched Uranium (LEU) sulfate solutions. Therefore, the determination of uranium concentration in uranium sulfate solutions is an important consideration for SHINE. Utilizing the Beer-Lambert Law allows for uranium concentration analysis using the characteristic visible absorption spectra of the uranyl cation dissolved in 1 mol/L sulfuric acid. The molar absorptivity (ε) at λmax was calculated previously using accurate standard solutions, with different UV-vis cells tested, impurities screened and data collection methods refined. However, comparison with the Davis-Gray uranium analysis technique revealed the importance of temperature control for accurate and precise measurement, with ε varying significantly as a function of temperature. To address this issue a newly installed Peltier unit was used to control cell temperature and ε determined from freshly prepared standard solutions over a range of uranium concentrations at a fixed temperature. To test the accuracy and precision of the refined technique, the concentration of uranium of 10 unknown samples were then measured at the same temperature. Finally, the impact of temperature on ε was analyzed for uranium, not just in 1 mol/L sulfuric acid but also 1 mol/L nitric and hydrochloric acids, thus facilitating an evaluation of technique general applicability.

478582 Screenning on Xylitol Production with Different Colombian Native Yeasts

Abstract:

Screenning of xylose consumption and xylitol production by Colombian native yeasts

Corresponding autor: jntorresl@unal.edu.co

Jessica Torres-Leiva[a], Sebastián Lancheros-Castañeda[a], Yina Cifuentes-Triana[a], Daniel Leal-Moreno[a], Juan Pablo Ortíz-Rosas[a], Mario Velásquez-Lozano[a]

[a] Universidad Nacional de Colombia - Bogotá, Department of Chemical and Environmental Engineering, Avenida Carrera 30 No. 45 - 03, 111321, Bogotá, Colombia.

Recently, bioconversion of biomass residual feedstocks has received particular attention because of its practical applications in various agro-industrial processes, as fuels and add-value chemicals production, energy generation and effluent remediation. Different agricultural residues, such as corn fiber, corn stover, wheat straw, rice straw and sugarcane bagasse, contain about 20 – 40% hemicellulose, the second most abundant polysaccharide in nature.

Hemicelluloses are heterogeneous polymers of pentoses (xylose, arabinose), hexoses and sugar acids. D-xylose is the principal sugar constituent of hemicelluloses and it can be reduced to different products as ethanol, xylitol, 2-3 butanediol and lactic acid through bioprocessing with microbial strains. Specifically, xylitol production has attracted much attention due to its potential use as natural food sweetener and sugar substitute for diabetics with anticarcinogenic properties. Xylitol bioconversion is limited to strains with the ability to metabolize xylose as carbon source.

Xylose assimilation ability was screened in eighty-five wild yeasts isolated from a Colombian distillery in a solid media with xylose as sole carbon source. Forty strains were able to growth in Yeast Medium Agar containing 20 g/L glucose, 20 g/L peptone, 10 g/L yeast extract and 20 g/L agar-agar at 48 hours. Xylitol production was evaluated on the xylose-assimilating strains under oxygen-limited conditions, 150 rpm. 30°C, and a media containing 30 g/L xylose, 1.7 g/L yeast nitrogen base (YNB, Sigma Aldrich®) and 5 g/L $(NH_4)_2SO_4$. Most of the strains were able to metabolize xylitol in 96h. Six strains of different species presentend the higher concentrations of xylitol: ***Candida tropicalis, Issatchenkia orientalis, Meyerozyma caribbica and Torulaspora delbruekii*** and two strains with no identity determined.

The best results were gained with ***Candida Tropicalis;*** a 4 g/L concentration of xylitol was obtained at 96 hours in 125 mL flasks with 50 mL of fermentation media. This strains fo were obtained from a Colombian distillery, particularly from bagasse in the floor near to a boiler, which suggests that ***Candida Tropicalis*** may be able to grow in a hydrolysate of sugarcane bagasse.

478616 Determination of the Best Conditions of Biodiesel Production with Free Form Lipases on Refining Palm Oil to Laboratory Scale

Abstract:

The enzymatic production of biodiesel has different opportunities and advantages over chemical production as improvements in the production process of biodiesel and removing byproducts, treatment of different raw materials, reduced costs associated with the operation [1,2,3].However the production process of biodiesel with enzymes in free form, must overcome different constraints by conditions in the medium as the stability of reaction, yields, the possibility of reuse or costs associated with the reaction[3]. This project was aimed to find the best conditions that would allow a high conversion to biodiesel in the laboratory, which was performed by enzymatic catalysis RBD palm oil with the formulation Eversa Transform feeding regularly with methanol to a reaction period of 9 hours. The best conditions biodiesel production were molar ratio of alcohol - oil 4.1, 34 ° C and percentage of enzyme 1.1% for a 74% with a achieved prediction by the model of 80%, where finally the biodiesel convertion is affected with increased alcohol in the medium and decreasing temperature, and linearly favored by amount of enzyme. These conditions were found by implementing the methodology MSR.

KEYWORDS: Biodiesel, Palm Oil, Lipases, Methanol

REFERENCES

1. Shweta Shah, Shweta Sharma and MN Gupta. Enzymatic transesterification for Biodiésel production. Indian Journal of Biochemistry & Biophysics. 2003; Vol. 40. 392 - 399.
2. Poppe, J. K., Fernandez-Lafuente, R., Rodrigues, R. C., & Ayub, M. A. Z. Enzymatic reactors for biodiésel synthesis: Present status and future prospects. ***Biotechnology Advances.*** 2015 ***33***(5), 511–525. http://doi.org/http://dx.doi.org/10.1016/j.biotechadv.2015.01.011
3. Fjerbaek, L., Christensen, K. V., & Norddahl, B. A review of the current state of biodiésel production using enzymatic transesterification. ***Biotechnology and Bioengineering***. 2009. ***102***(5), 1298–1315. http://doi.org/10.1002/bit.22256

478617 The Formation of Vesicles through the Modification of Tyrosine-Derived ABA Tri-Block Copolymer Membranes

Abstract:

Amphiphilic ABA triblock copolymers are studied as a promising biomimetic substitute for lipids that form vesicles, a nano-scale morphology that is used as a carrier for drug delivery. The hydrodynamics of the tyrosine-derived PEG-*b*-oligo(desaminotyrosyl tyrosine octyl ester-suberate)-*b*-PEG (PEG-oligo(DTO-SA)-PEG) was investigated using dissipative particle dynamics (DPD) simulation techniques. This technique is a coarse-grained (CG) molecular dynamics-based approach that uses soft repulsive interactions between beads that represent clusters of atoms or molecules. Computational results have shown that this polymer membrane forms a disk-shaped micelle only. Therefore, by inserting polymers that are folded 180 degrees throughout the membrane, we have modified the polymer membrane so that other morphologies are formed, namely vesicles. Varying chain lengths of both hydrophobic and hydrophilic segments have been shown to demonstrate different morphologies of this ABA triblock copolymer. For chain length combinations that form vesicles, the radius of the vesicle and the angle distribution of the polymers were computationally calculated and graphed based on the trajectory files from the simulations. These results show a strong linear relationship between the chain lengths and vesicle size. We also propose potential experimental applications of these results, which could demonstrate that these simulations can be observed in a laboratory setting.

478618 Characterizing Mammalian Cell Shear Sensitivity to Improve Cell Culture Bioreactor Efficiency

Abstract:

Large scale bioreactors are used by the Biotechnology Industry for the production of therapeutic proteins. Bioreactors incorporate gas sparging to provide oxygen requirements for cell cultures, and facilitate mixing via an impeller. Agitation from sparging generates bubbles, which can damage cell cultures at the media/air interface within the culture broth and at the interface of the culture broth and tank headspace. Cell damage can impact VCD (viable cell density) and Culture Viability. To mitigate these challenges, Nonionic surfactants are included within the cell culture media. Specifically, surfactants known as Poloxamer, were evaluated in this study. Poloxamers allow for hydrophobic interactions with cell membranes, and hydrophilic interactions which sterically hinder cell-bubble attachment. This disruption of cell-bubble attachment reduces cell death in heads of foam. These interactions reduce cell damage to improve bioreactor performance. Characterization of Poloxamer effectiveness at protecting cell cultures during bioreactor operations can provide significant benefits to the manufacturing of therapeutic proteins. To evaluate the effectiveness of Poloxamer, a motorized cylindrical impeller was used to shear cells in a media solution containing Poloxamer. Experiments were designed and executed to investigate Poloxamer concentration, shear rate (RPM), and culture viability. Samples were analyzed via trypan blue dye exclusion to measure VCD and culture viability. Results from this study demonstrate a reliable method to quantify cell shear sensitivity in the presence of different Poloxamers

478623 Development of Nanoparticle Encapsulation for Targeted Drug Delivery

Abstract:

Polymer nanoparticles are commonly used in the research lab as a method to controllably release a drug. However, the high surface to volume ratio of the nanoparticles results in the drug release commonly taking place over the course of 24 to 48 h. For longer term delivery, other methods need to be developed. We have developed a method to create injectable hydrogels filled with polymer nanoparticles. The hydrogel acts as a barrier to slow down diffusion of the drug from the nanoparticles effectively extending release of the drug.

Polylactide-co-glycolide polymer nanoparticles loaded with a model drug were prepared by impingement jets mixing and removal of the solvent using an ultrafiltration stir cell. Sucrose was added to the aqueous nanoparticle solution and the mixture was freeze dried to produce polymer nanoparticle powder. The nanoparticles were loaded into an injectable hydrogel formulation to modulate the controlled release of a model drug. The drug release kinetics of the polymer nanoparticle/hydrogel composite were evaluated *in vitro* to determine the effect of adding a mass transfer barrier on the rate of drug release from polymer nanoparticles. The use of a hydrogel effectively increased the drug release time from 2 d to 12 d.

478712 The Effect of Poly (beta amino ester) and Dopamine Coatings on the Surface Activity of Iron Oxide Nanoparticles

Abstract:

This study aimed to determine if poly (beta-amino ester) (PBAE) and dopamine (DA) coated iron oxide nanoparticles (IONPs) would impact the generation of reactive oxygen species (ROS) in the presence of an alternating magnetic field (AMF). IONPs were synthesized by a co-precipitation method and were coated with PBAE polymers containing different amounts of dopamine and dopamine alone. The molar concentration of dopamine was altered in each PBAE polymer while the ratios of PEG400DA:DEGDA (2:1) and the diacrylate:amine (1.2:1) were held constant. PBAE 0.1% DA and PBAE No DA were successfully synthesized and compared to previously synthesized PBAEs with 1% and 10% Dopamine concentration. PBAE No DA and PBAE 0.1% DA were both successful in coating and degrading from the IONPs. The results of the methylene blue decolorization assay with AMF exposure demonstrate that coated IONPs show more evidence of ROS generation than uncoated or degraded IONPs, and Dopamine coated IONPs completely inhibited ROS generation. PBAE No DA coated IONPs generated ROS most effectively therefore suggests that dopamine causes too strong binding of PBAEs to IONPs and compounds with less reactive anchoring groups should be studied for alternative PBAE polymers.

Research Supported by the National Science Foundation REU Program #EED-1460486

478723 Optimization of a NG-Hydroxy-L-Arginine Electrochemical Sensor

Abstract:

NG-Hydroxy-L-arginine, NOHA, is a suspected biomarker in rare and aggressive cases of breast cancer (1). NOHA was found to have antiproliferative and apoptotic actions on arginase expressing human breast cancer cells (1). The most promising approach in the treatment of cancer is to undergo rapid treatment options as a result from early diagnosis of cancer (2). Detecting the onset of cancer in a fast, cost efficient, and at optimal range is essential for greater treatment options. Electrochemical detection is a proven, established, cost-effective method to successfully detect low levels of analyte concentrations. Ideally one wants to detect NOHA on a nanomolar scale, because NOHA is present in serum in low concentrations. The goal of initial experiments where to optimize a NOHA sensor at a bare glassy carbon electrode. By comparing the detection methods of cyclic voltammetry to differential pulse voltammetry, we evaluated the best detection techniques for future experiments in NOHA detection. Each method will be evaluated on several criteria. First, we will monitor the oxidation peak for any voltage shifts as a function of concentration. Our experience indicates that a shift in the peak voltage would make the creation of a precise calibration curve difficult. Next, we will observe the relationship between the current response and concentration. The most accurate calibration curves are linear, and we will report on the linearity of NOHA's oxidation current response versus concentration. In the future attachment of copper polyaniline film will be used in enhance the surface of the electrode using the best electrochemical detection method.

(1) R. Singh, S. Pervin, a Karimi, S. Cederbaum, and G. Chaudhuri, "Arginase activity in human breast cancer cell lines: N(omega)-hydroxy-L-arginine selectively inhibits cell proliferation and induces apoptosis in MDA-MB-468 cells," Cancer Res., vol. 60, no. 12, pp. 3305–3312, (2000)

(2) Zhen Zhang, Robert C. Bast, Jr., Yinhua Yu, Jinong Li, Lori J. Sokoll, Alex J. Rai, Jason M. Rosenzweig, Bonnie Cameron, Young Y. Wang, Xiao-Ying Meng, Andrew Berchuck, Carolien van Haaften-Day, and

Neville F. Hacker, Henk W. A. de Bruijn, Ate G. J. van der Zee, Ian J. Jacobs, Eric T. Fung, and D. W. Chan, "Three Biomarkers Identified from Serum Proteomic Analysis for the Detection of Early Stage Ovarian Cancer," Cancer Res., vol. 64, pp. 5882–5890, (2004)

478725 Pentablock Copolymer Mediated Crispr/Cas9 Delivery Strategies in Astrocyte Type Cells Transdifferentiated from Mesenchymal Stem Cells for in Vitro Gene Therapy

Abstract:

Parkinson's and Alzheimer's are the most common neurodegenerative diseases in United States. Currently, these neurodegenerative disease can't be cured, and medications are only used to improve symptoms and treat patients. The morphological hallmarks of inflammatory and neurodegenerative disease are caused by enlargement of astrocytes and accumulation of macrophages. In recent literature, production of granulocyte-macrophage-colony–stimulating factor (GM-CSF) by astrocytes has been suggested to contribute to the macrophage response. GM-CSF secreted by astrocyte has shown to enhance the local inflammatory response to central nervous system injury and to activate macrophage aggregation. Also, increasing evidence suggests that α-synuclein aggregates can activate astrocyte cells to induce these neruroinflammation. In this study, amphilic pentablock copolymer was used to efficiently deliver gene editing tool of CRISPR/Cas9. CRISPR/Cas9 was used to efficiently target GMC-SF gene to decrease the gene's effects on Astrocyte that causes astroctytosis, which is a hallmark of various neurological injuries an in inflammatory processes. Polymer transfected with CRISPR/Cas9 on GDNF derived Astrocytes showed lower GM-CSF concentration compared with just CRISPR/Cas9 delivered. Additionaly, it is shown that primary astrocytes loaded with α-synuclein secretes higher concentration of GM-CSF compared with cells that was treated with CRISPR/Cas9. With this study, we were able to show that the amphiphilic polymer is effective for delivering CRISPR-Cas9 gene editing tool to treat genetic diseases.

478733 3D Printed Microfluidic Platform for Non-Invasive Smart-Agriculture Sample Collection

Abstract:

The goal of this project is o advance the development and capacity of hydroponic â€œurban growingâ€ smart-crops by developing a low-cost, minimally-invasive monitoring device to provide feedback for nutritional regulation. However, conventional analytical techniques are inherently destructive and sacrificial in nature and cannot be conducted constantly, limiting the amount of obtainable feedback and response when optimizing and plant growth and yield. We propose a open system microfluidic recirculation concept with a droplet interface to form contact with plant fluids to establish equilibrium between a plant's natural fluids and an analytical device and analyze changes in microchip fluid concentrations to serve as an indicator to the changes in the plant fluid concentration levels.

478632 Self-Assembling paclitaxel Nanofibers for Controlled Drug Release

Abstract:

Peptide-drug conjugations have been of great interest in recent years for cancer chemotherapy. The 3D self-assembly process of peptide hydrogel is governed by many different types of forces, which we can carefully manipulate to control the pharmacokinetics of the drug. However, the critical role of hydrophobicity and π-π interaction in the morphology has not been fully explored. Here we have developed monodisperse drug amphiphiles that self-assemble into helical super-coiled filamentous structures. We conjugated the hydrophobic drug Paclitaxel to the peptide sequence FFFRGDR via a biodegradable disulfylbutyrate (buSS) linker. The peptide sequence incorporates several aromatic amino acids Phenylalanine and the tumor penetrating peptide RGD. The π-π interaction and steric effect causes torsional strain in the β-sheets and thus the right hand twisting, leading to the "super coil" nanofilaments. These helical nanofibers have been shown to have enhanced stability compared to straight peptide nanofibers as shown by a lower critical micellization concentration (CMC). They have a slower and more sustained release of the drug content. This suggests possibility for less frequent dosing due to the slower release and smaller fluctuation in the concentration of the drug, which is beneficial to the patients. We believe this strategy can be extended to fabricate morphologies and stability of nanostructures with π-π interactions and hydrophobic interactions. These nanofilaments with various structures and mechanical properties can be used for hydrogel formation with different anti-cancer drug release rates, leading to new opportunities for the local treatment and drug delivery for multiple cancer chemotherapy.

478633 Albumin Extraction Process from Bovine Blood SERUM

Abstract:

The albumin is a protein present in blood that helps transporting different substances through the body. It is also in charge of keeping the required oncotic pressure in the vascular system. This protein has a half-life of 18 days and it can be obtained from the albumen of eggs. Industrial production of Albumin is of major interest because it can be used for hemodynamic support in patients with severe sepsis and septic shocks, and also due to its potential to become a major monomer in 3D food printing.

Due to different diseases associated with poultry derivatives consumption (salmonella, avian influenza, antibiotics content, etc.), egg-extracted albumin might represent health risks. In this sense, alternative sources of albumin are needed for industrial exploitation. Recently, it has been stated that slaughterhouse blood can be used as raw material for bioactive compounds. So, in this work we used bovine serum blood as albumin source, which can be obtained as a waste stream from slaughterhouses.

The content of blood in animals is approximately 3% of the total weight. According with production reports in Colombia, each sacrificed bovine produces an average 12,5 kg of blood. Correspondingly, bovine blood serum contains 4.7% of albumin in a wet basis. Table 1 show an approximately composition of the blood plasm.

Table 1. **Composition of the blood plasm**

Compound	Percentage	
Water	94.50%	
Protein	7%	
	Albumin	4%-5%
	Globulin	2%-2,5%
	Fibrinogen	0,3%-0,4%
Others	3%	

The price of the albumin is calculated according the protein grade. Samples analyzed in different Colombian slaughterhouses showed a protein content 73% HSC, resulting in a price of US$ 0.38 per kilo. This commercial opportunity was the driving for developing a standardized a bovine albumin extraction process.

The developed process is described in *Figure 1*. First, as-received bovine blood was stabilized by adding an anticoagulant to avoid fibrin clots formation. The stabilized solution was centrifuged obtaining blood serum and hemoglobin. The blood serum was subsequently subjected to a pH controlled extraction and centrifugation as described in *Figure 2*. Albumin productivity with respect to bovine blood was about 1, 5% wt, and the extraction efficiency was around 70% with respect the total amount of albumin in the blood serum. The obtained product was a white solid that settled in the form of fine sheets.

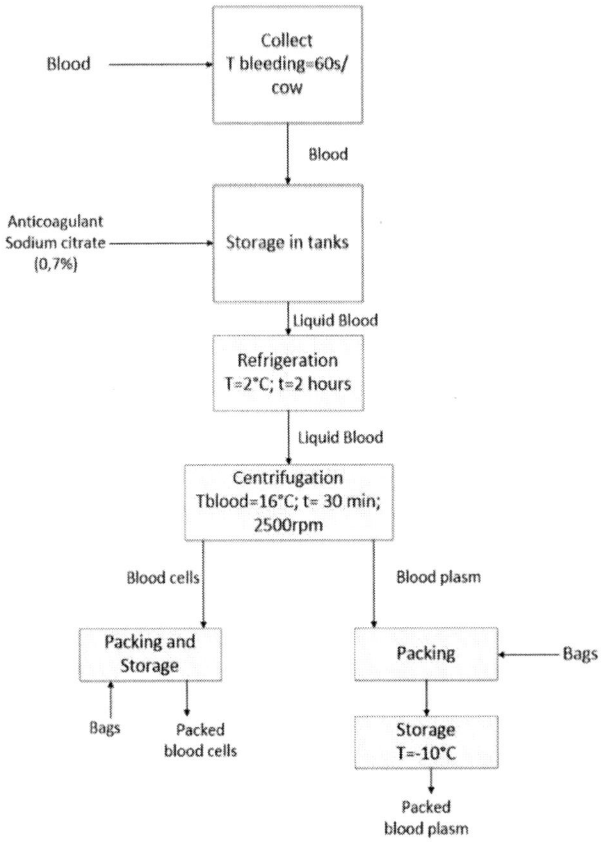

Figure 1. Bovine blood serum extracting process from bovine blood

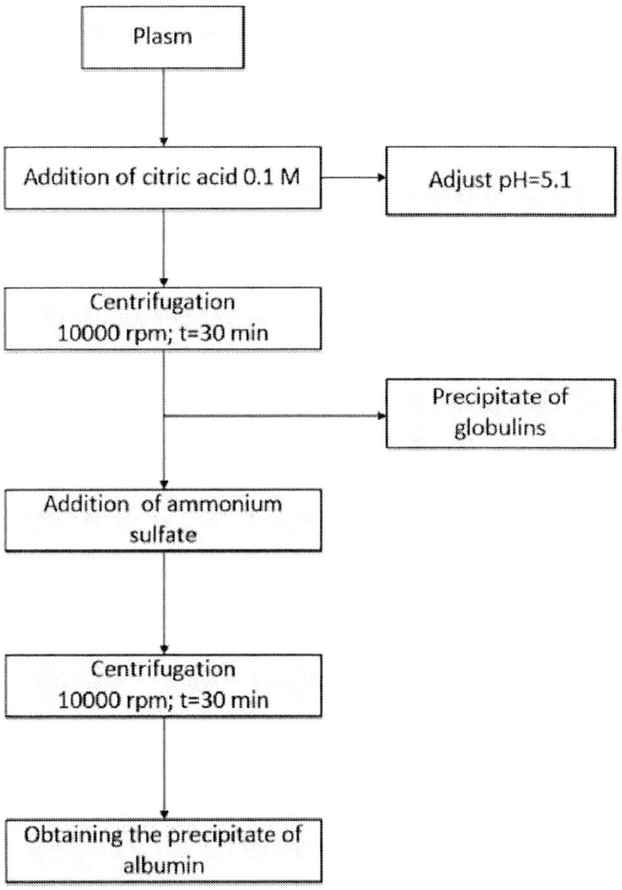

Figure 2. Albumin extracting process from plasm

478701 Chondrogenic Differentiation Potential of Human Adipose Stem Cell Spheroid Forming on Glycol Chitosan and Gelatin Hydrogel

Abstract:

Cartilage tissue often suffers from difficulties to repair after injury due to a lack of blood vessel and neural systems. Furthermore, since cartilage tissues serve as buffers between bones and are often located at joints, irreversible injury or dysfunction of the whole joint might occur if the damage is not repaired in time. As a result, repairing cartilage tissue has long been an essential topic in tissue engineering.

Human adipose derived stem cells (hASCs) are proved to have the potential to differentiate into different lineage including adipogenic, osteogenic and chondrogenic lineage. They are considered to be the cell sources in cartilage tissue engineering.

Glycol chitosan, which is recently studied for cartilage repairing due to its similar structures to glycosaminoglycans found in cartilage and the decent biocompatibility and degradability in vivo. In this study, hydrogel mixed with glycol chitosan and gelatine is used as a carrier of hASCs and abilities for the cells to adhere, proliferate and differentiate on the hydrogel composite are evaluated, in order to analyse the potential of chodrongenic differentiation of glycol chitosan.

The existence of gelatine can enhance the cell adhesion ability of the composite, however, as gelatine is released out of the composite with time, hASCs will start to form a spheroid structure. The formation of spheroids was attributed to the releasing of gelatine and the characteristic of glycol chitosan, which the surface of glycol chitosan will induce the formation of cell spheroids, in which researches have indicated that cell spheroids can stimulate the differentiate ability.

These spheroids were observed by fluorescence and SEM to test their properties and gene expression of differentiation. Further stimulation using chodrongenic differentiation medium to trigger the spheroids to differentiate was conducted, the results showed chondrogenic differentiation performance appeared in cell spheroids. Moreover, the spheroids formed with the induction medium were larger and had more complete shape of sphere. They also had higher performance in chondrogenic differentiation.

478772 Characterization of an Electrostatic-Based Air Sampler for the Capture of Airborne Norovirus

Abstract:

The need for an air sampler has been in demand, as airborne transmitted viruses like Norovirus, also known as the winter-vomiting disease, produces 685 million cases a year around the world1. The principal of electrostatic precipitation (ESP), although well-known for air filtration, has been recently used as a novel method for sampling for airborne viruses. The ESSA (ElectroStatic Sampler for Air) is a novel point-of-care device developed in-house for the capture of airborne virus. Two designs were created for the ESSA, that vary the needle organization; the cross design and the circular design. This project focuses on evaluating the performance of these two ESSA design for capturing airborne viruses. Based on the characterization results, the circular ESSA has the highest particle count efficiency while the cross ESSA has the best current versus voltage relationship.

478827 Dynamic Tissue Culture Model for Ex Vivo Recreation of the Mechanical Microenvironment of the Lung

Abstract:

The purpose of the research is to create a lung-on-a-chip device that simulates the mechanical microenvironment. Traditional cell culture methods fail to account for the effects of mechanical factors, such as strain and pressure, on lung cells. Because lung cells respond to mechanical factors as well as biochemical factors, a failure to introduce mechanical factors in the microenvironment of the cells results in inaccurate or incomplete results. The goal of this research is to create a dynamic model for cell culture and assess its performance when compared to traditional cell culture methods. The dynamic model provides researchers with critical information that the traditional culture method cannot supply, such as the behavior of a mechanosensitive protein under in vivo conditions. This new model will be used to study mechanosensitive channel defects that cause cystic fibrosis.

Research Supported by the National Science Foundation REU Program #EED-1460486

478971 Thermally Triggered Drug Release from Injectable Hydrogel Composite for Local Chemotherapy Delivery to Tumors

Abstract:

A drug delivery system was designed to induce the controlled release of encapsulated anticancer agents in response to elevated temperatures using thermosensitive liposomes suspended in a thermally sensitive polymer, poly(N-isopropylacrylamide)-graft-chondrotin sulfate. Drug release is triggered by a heated environment at the temperature range of cancer tissue hyperthermia (T > 40°C). Heat and anticancer agents were utilized for this system because their combination is believed to have a synergistic cytotoxic effect on internal cancerous tissue. Two types of liposomes were produced using 1,2-dipalmitoyl-sn-glycero-3-phosphocholine (DPPC) and 1,2-distearoyl-*sn*-glycero-3-phosphocholine (DSPC), which have transition temperatures of 41°C and 50°C, respectively. The lipid bilayers of the liposomes melt above these temperatures, which allows the release of the entrapped drug. The main objective of this study was to determine if the release of a model molecule, calcein, from the liposome-hydrogel composite occurs in a pulsatile manner in response to multiple spikes in temperature. Three types of liposome-hydrogel composites were used, which consisted of DPPC liposomes, DSPC liposomes, and a 50/50 mixture of DPPC and DSPC liposomes. The composites were maintained in phosphate buffered saline (PBS) and incubated *in vitro* at physiological temperature, 37°C. Every 24 hours, the samples were exposed to a temperature spike lasting 30 minutes, with the first two spikes at 41 °C and the last two at 50 °C. The amount of calcein released during each temperature spike was quantified with fluorescence spectroscopy and normalized to the total loaded calcein. The composite containing only DPPC liposomes released calcein after the first temperature spike with the ability for continued release with each subsequence spike. The composite containing only DSPC liposomes had very little calcein release at the 41°C spikes, and released calcein at a greater rate of release during the 50°C spikes. The composite containing both liposome types was able to release calcein equally at both temperatures but not to the same intensity as the other two samples types. Overall, the data indicate the potential for heat-triggered release of chemotherapy drug capable of repeated release after multiple temperature spikes. These results are significant because they suggest that this injectable hydrogel has potential for clinical relevance. When implanted within a tumor, the composites can release chemotherapy drugs in a pulsatile manner in response to hyperthermia. The combination of lipid types provide a greater temperature control to extend the treatment time. Future work will focus on characterizing the thermal properties of the hydrogel to have the composite be heated internally by ferromagnetic nanoparticles.

479002 Biodiagnostic Assay Using Multibranched Gold Nanoantennas for Ultra-Sensitive Biomarker Detection Via Sandwich Architecture

Abstract:

This system utilizes the plasmon enhancement capabilities of multi-branched gold nanoantennas (MGNs) to achieve ultra sensitive and specific detection of Prostate Specific Antigen (PSA) via Surface Enhanced Raman Spectroscopy (SERS). This extremely low detection limit of PSA was achieved with a biodiagnostic sandwich architecture consisting of a glass substrate coated with anti:PSA MGNs along with a top MGN probe functionalized with an additional anti:PSA and Raman tag for recognition. One early indicator of prostate cancer is the elevation of PSA levels in the blood due to tumor formation in the region. By improving the detection limit for this PSA biomarker, the early onset of this disease can be screened and subsequently treated early in its progression. In order to determine the PSA detection limit of the assay, different concentrations of the antigen were incorporated and then analyzed via SERS. A detection limit of 5 fg/mL PSA was achieved. In addition to these PSA sensitivity trials, a non-specific protein, immunoglobulin, was analyzed to test the specificity of this assay. These experiments demonstrate the biomarker detection capabilities and provide a platform for future detection of other cancer biomarkers.

479054 Coarse-Grained Simulations of the Self-Assembly of Skin-Relevant Lipid Structures

Abstract:

The barrier properties of the skin are largely governed by the lipid lamellae of the outermost layer of skin, the stratum corneum (SC). While the composition of the SC is known, the molecular-level features of the lipid organization are not. This information would be useful for repairing an impaired skin barrier or selectively bypassing the barrier, e.g., for transdermal drug delivery. Molecular simulation allows precise control over system composition and direct visualization of the system structure, making it a useful tool for studying such systems. The slow kinetics of lamellar self-assembly make using atomistic models computationally inefficient. Thus, coarse-grained models, where groups of atoms are treated as individual interaction sites, are employed to study the self-assembly of SC lipids. Here, coarse-grained models of SC lipids, including CER NS C16, CER NS C24, cholesterol, and free fatty acids, are simulated to gain insight into the low-energy structures adopted by mixtures of these lipids. Self-assembly of both single and stacked bilayers in solution is studied by cooling systems from a high-temperature, disordered state. Structural properties, such as lipid positioning, angle between lipid tails, tilt angle of lipid tails, nematic order parameter, and bilayer lipid density are calculated. These properties are compared between simulations to determine the effects of lipid concentration on bilayer structure. This research increases understanding of the structural role of various lipids in SC organization and is an important step in the use of accurate coarse-grained models of the SC lipids.

479089 Identifying Differential Gene Expression in Cytotoxic T Cells Induced By B16F0 Melanoma Exosomes

Abstract:

Recent clinical studies demonstrate how engaging a patientâ€™s own immune system can be an effective tool in the fight against cancer. Yet, a barrier for broadening the clinical benefit of immunotherapy is identifying how tumors suppress the ability of immune cells to kill malignant cells. An important emerging mode for cellular cross talk is exosomes, which are nanoscaled particles secreted by tumors that carry proteins and RNA. To better understand how tumor-derived exosomes can alter immune cell function, this study investigated the effects of tumor exosomes on cytotoxic T cells. Specifically, we quantified how exosomes from B16F0 melanoma cells altered gene expression over time within CTLL2 cells, a mouse cytotoxic T cell line. Changes in mRNA over time was analyzed by RNA sequencing. Pathway enrichment algorithms were used to identify how B16F0-derived exosomes alter the function of cytotoxic T cells.

479097 Development of a Microinjection Platform for Chlamydomonas Reinhardtii

Abstract:

The ability of the microalga *Chlamydomonas reinhardti* to produce high-value chemicals in a mild environment makes this species conducive towards low-environmental impact production practices [1-2]. Genetic manipulation of the unicellular organism to produce such desired molecules has been practiced since the late 1980s [3]. While there currently exist several functional transformation methods, first generation expression rates remain low, peaking around 60% in laboratory environments [4]. A microinjection platform to deliver exogenes into the cell may produce significantly higher expression rates whilst also providing a simplified transformation method which does not necessitate the removal of algal cell walls as required in many current methods [5]. This study describes the initial development of such a device.

The microinjection platform for this study uses a wide-diameter pipette with mild suction to immobilize a cell-wall-intact microalga, then impaling the cell with a second, smaller-diameter pipette filled with genetic material. Implicit with microinjection is a breach of the cell wall and membrane which reduces a cellâ€™s viability by exposing the intracellular environment to the extracellular impalement medium. In this study, multiple parameters were controlled and varied to optimize the survival rate of cells during and after impalement. These parameters included the stage of cell growth during which impalement took place, the diameter of the injection pipette, and the medium in which the cells were impaled. Further development of this microinjection platform will include affirming the integrity of the injection pipette after multiple impalements, determining the optimal duration of an impalement, optimizing the voltage used to electroosmotically deliver material into the cell, and developing a medium to be injected into the cells.

References

[1] S. P. Mayfield, et al., "Chlamydomonas reinhardtii chloroplasts as protein factories," Current Opinion in Biotechnology, 18, 126-133 (2007).

[2] C. Griesbeck, et al., "Chlamydomonas reinhardtii: A protein expression system for pharmaceutical and biotechnological proteins," Molecular Biotechnology, 34, 213-223 (2006).

[3] J. M. Coll, "Review. Methodologies for transferring DNA into eukaryotic microalgae," Spanish Journal of Agricultural Research, 4, 316-330 (2006).

[4] A. Q. Rao, et al., "The myth of plant transformation," Biotechnology Advances, 27, 753-760 (2009).

[5] A. Crossway and D. Facciotti, "Plant cell microinjection technique". US Patent 4743548, 10 May 1988.

479149 Incorporation of Benzalkonium Chloride to Improve the Antimicrobial Properties of Lignocellulosic Material

Abstract:

TEMPO nanofibrillated cellulose (TNFC), carboxymethyl cellulose (CMC), and lignin were applied as nanoscale carrier molecules for benzalkonium chloride (BZK), a well-known organic biocide. The resulting hybrid suspension was analyzed for biocidal activity against *Escherichia coli* (*E. coli*) and dried for scanning electron microscopy analysis using a solvent exchange and freeze drying technique. The results indicate that there is slow release of BZK in suspension, which is desirable for long-term antimicrobial activity. Instant biocidal activity can be achieved with a minimal inhibitory concentration of 0.116 M BZK-TNFC hybrid suspension. Lignin and CMC used in complex with BZK to create a 0.046 M BZK-Lignin and 0.046 M BZK-CMC suspension demonstrated increased antimicrobial activity with longer exposure times. Initial concentrations of E. coli were 3.1×10^{-6} and 0.5×10^{-6}, while 24 hour exposure at 37° C resulted in complete inhibition of bacteria.

479215 An Image Analysis Tool for Mapping Cilia-Driven Particle Crystallization

Abstract:

An Image Analysis Tool for Mapping Cilia-Driven Particle Crystallization

Aiello, Joseph M., Markovetz, Matthew R., Parker, Robert S.

Department of Chemical and Petroleum Engineering, University of Pittsburgh

An important aspect of particle-based drug delivery is the degree of crystallinity of the delivered particles *in vivo*. Brownian motion affects particles in a fluid under 6 microns in diameter enable the formation of crystalline structures. Larger particles require that a driving force be applied to them. In the case where particles are inhaled into human lungs, cilia within the lungs may impart such a force on the particles in an effort to clear them from the airways. In this work, images were taken using confocal microscopy of particles on well-differentiated, ciliated human bronchial epithelial cell cultures. Stacked 2D image files of the particles were generated post-assembly. Using an augmented form of the Circular Hough Transform function, each stack of particles was evaluated and particle centers were identified. From the data obtained during each stack evaluation, the overall crystal structure in this 3D system can be determined.

478839 Isolation and Identification of Hypoglycemic Agents from the Costus Family of Plants

Abstract:

The tropical plants of the Costus genus are abundant in the Caribbean area, especially in Puerto Rico. â€œInsulinaâ€ or the insulin plant is one of them (Costus speciosus, Costus spicatus, Costus spiralis, etc.) and a common practice in folk medicine is to prepare tea from its leaves and use it to reduce high blood-sugar levels (hypoglycemic agent). Identifying and separating a pharmacophore or several active ingredients of the insulin plant is the goal to achieve. A Soxhlet extraction is the method to be used in order to generate solutions containing the plant leavesâ€™ essentials. Six (6) solvents (500 mL for each) with variant polarity are used for the extraction, in addition to acidic and basic aqueous solutions were also used. A weighed amount (100 g) of the leaves were cut into 2.54 cm (1 in) squares and placed in the Soxhlet extractor together with the 500 mL of solvent, heated to a boil in order to conduct the continuous extraction for a total of six (6) hours. The resulting extracts were analyzed by spectroscopic techniques (1H NMR, and 13C NMR) in order to determine functional groups and/or structural features present in the crude extract. Preliminary results from this spectroscopic analysis have indicated the presence, among others, of known compounds such as essential oils, proteins, and antioxidant components. Infrared spectroscopy will be conducted in order to find the presence of other known hypoglycemic agents such as sulfonylureas and corosolic acid (glucosol).

478884 Revealing Ultrafast Structural Motions of the GFP Model Chromophore in Solution

Abstract:

Fluorescent proteins as biomarkers have provided decades of valuable research insights for the scientific and engineering community. The green fluorescent protein (GFP) has been a favorite for bioimaging due to its bright fluorescence and high quantum yield. Its success has inspired bioengineers to modify the chromophoreâ€™s structure and environment to create many new fluorescent proteins with better properties. Despite the large success of engineered GFP derivatives, ongoing research seeks to solve some of the drawbacks of current fluorescent proteins, including low brightness, fast photodegradation, and poor contrast. Critical to this research is knowledge about the interplay between chromophore fluorescence and local environment, a paradigm that is still not fully understood for GFP. Notably, the specific excited state structural motions preventing the GFP model chromophore 4-hydroxybenzylidene-1,2-dimethylimidazolinone (HBDI) from significantly fluorescing when taken out of the protein pocket are undetermined.

Femtosecond stimulated Raman spectroscopy (FSRS) in conjunction with transient absorption (TA) spectroscopy provides critical insights into the structural motions that compete with fluorescence when HBDI is in solution. Analysis of intensity and frequency dynamics reveals that the molecule undergoes internal conversion by evolving into a twisted internal charge transfer (TICT) state on a ~2.0 ps timescale from an intermediate emerging after a ~500 fs vibrational cooling component. To glean further structural information from the data, we innovated a new 2D Raman approach for comprehensive visualization of anharmonically coupled vibrations. Using this method, we uncover a 232 cm^{-1} out-of-plane bending mode modulating a 867 cm^{-1} phenol hydrogen out-of-plane (HOOP) mode as HBDI evolves into the TICT state, providing compelling evidence for the functional relevance of conformational twisting as the molecule approaches the conical intersection.

The interplay between chromophore environment and fluorescence is further revealed by studying the effects of adding a viscous glycerol solvent. In contrast to that in pure water, HBDI in glycerol exhibits longer excited state dynamics, echoing the increased excited state lifetimes of the molecule within the protein pocket. Surprisingly, the 232 cm^{-1} vibrational modulation disappears upon adding glycerol; instead, a 125 cm^{-1} weak out-of-plane mode and a 275 cm^{-1} in-plane methyl rocking mode appear. This change in structural dynamics is likely because the energy barrier for the 232 cm^{-1} out-of-plane mode mode becomes prohibitively large in glycerol, whereas this is not the case for the two new modes. Differences in the vibrational coupling matrix due to a changing local environment show that increasing steric hindrance significantly inhibits the efficacy of nonradiative energy dissipation through the conical intersection for HBDI. In context of the GFP chromophore, the Î²-barrel of the protein pocket provides an intrinsic barrier to HBDI rotation that prevents conformational twisting as an alternative pathway to fluorescence. These findings reveal the molecular motions that govern the fate of the GFP chromophore in solution, and correlate the appearance of these specific motions with chromophore environment. New knowledge about the chromphoreâ€™s photophysical properties provides valuable information for the quest to engineer better fluorescent proteins to benefit the growing bioimaging community.

478890 Characterizing Cell-Free Metabolism through Physiochemical Optimization for Biosynthetic Pathway Engineering

Abstract:

Traditional *in vivo* methods for building genetic constructs, studying enzymes, and probing metabolism is slow and difficult. A platform is needed in order to rapidly and efficiently study metabolism, observe enzyme kinetics, and prototype pathways for industrially relevant chemical production. Crude lysate cell-free systems have been utilized to quickly produce chemicals through cell-free metabolic engineering (CFME), such as n-butanol via a 17-step pathway. By combining cell-free protein synthesis with metabolic engineering (CFPS-ME), we are able to rapidly monitor and manipulate physiochemical conditions in order to test metabolism and prototype enzymes. We describe a series of experiments outlining the differences in metabolic profiles of the two systems, the effects of alternative energy sources, and the ability of CFPS-ME for enzyme titrations in order to optimize pathways. We found that the components of CFPS reactions inhibit the production of n-butanol, and that phosphate-minimal and pH controlled setups are necessary. This is the first time that metabolite profiles have been extensively characterized in CFME, and this gives insight into how to expand these cell-free frameworks to accelerate the metabolic engineering design-build-test cycle.

479234 Enzyme-Based Conjugates Capable of Bacterial Decontamination

Abstract:

Surfaces in hospitals and medical clinics can easily become contaminated with bacterial species that can rapidly spread, potentially causing infections. This is a serious health concern that medical establishments need to combat on a daily basis. Despite efforts to reduce the spread of bacteria through sterilization methods with harsh chemicals, there is no guarantee that all bacteria are killed in real time. Herein, the development of a coating containing an active decontaminating species is proposed. The coating is based upon the integration of bio-nano conjugates, comprised of a system of enzymes bound to nanosupports, into paints. The enzymes, glucose oxidase and chloroperoxidase, were bound physically and covalently, respectively to multiwall carbon nanotubes to increase the coatingâ€™s functionality. The surface chemistry and morphology of the conjugates were observed as well as the loading and activity of the enzymes upon immobilization. Activity was reported with respect to the free enzyme via colorimetric assays. Preliminary results suggest that the conjugate system generates the active decontaminant (hypochlorous acid) in concentrations that would be suitable for bacterial decontamination.

479363 Biosynthesis and Purification of Nicotinamide Adenine Dinucleotide for Mechanistic Studies of Toxins

Abstract:

Nicotinamide adenine dinucleotide (NAD^+) has many key biological roles, including its function as an essential co-factor as well as acting as a signaling molecule. For these reasons, dysregulation of NAD^+ is characteristic in several key pathologies (e.g. colorectal cancer and necrotizing fasciitis). Moreover, it is not surprising that toxins such as the cholera toxin, diphtheria toxin, and tuberculosis necrotizing toxin utilize this activity to induce host cell death. Even while toxic beta-NAD glycohydrolases pose a serious threat, there is still much to be learned about their roles in pathogenesis. Only with a better understanding of how these enzymes function will it be possible to develop better diagnostic and therapeutic agents. In hopes of developing chemical biology tools to probe the role of beta-NAD^+ glycohydrolases, NAD^+ analogs with a wide range of pK_a values were biosynthesized and purified. With the natural substrate analogs in hand, it will be possible to better understand the enzyme catalyzed reaction, thereby providing a blueprint for future probe designs.

479402 Chondrogenesis of Encapsulated MSCs

Abstract:

Chondrogenesis of Encapsulated MSCs

Patrick Erickson, Kevin B. Miles, Howard W.T. Matthew

Abstract

The human body lacks a mechanism to regenerate damaged articular cartilage. Shallow cartilage defects do not heal, and may progress until much of the cartilage in the joint has degraded. This study explored a modular tissue engineering approach that could be used to regenerate articular cartilage. Rat bone marrow mesenchymal stem cells (MSCs) were encapsulated in hollow, spherical microcapsules composed of an ionic complex of chitosan and either hyaluronan (HA) or a mixture of HA and chondroitin-4-sulfate (C4S). The encapsulated MSCs were cultured in chondrogenic medium for 3 weeks to induce differentiation. Cell morphology and organization were evaluated by phase contrast microscopy, while viability was assessed by Calcein/Ethidium staining. Histological staining was used to characterize the distribution of secreted proteoglycans and collagens type I, II, and X. Total collagen was determined using a hydroxyproline assay. Results to date show that the MSCs survived in all capsule types. In all of the capsules with hollow interiors, the MSCs distributed themselves into small clusters on the interior capsule surface. When cultured in chondrogenic medium, these cell clusters grew into larger, rounder structures, which were likely to be chondrocytes along with their secreted extracellular matrix (ECM). Hollow capsules made of HA had the most numerous cell clusters, and after 3 weeks in chondrogenic medium, the cells and ECM filled most of the capsule interior. Hollow capsules made of a mixture of HA and C4S appeared to have holes in their membranes, and the cell clusters in chondrogenic medium became large and round, but remained as discrete clusters, unlike in the HA capsules. Capsules made of HA with an interior filled with collagen type I shrank upon formation and condensed all of their

cells together. The cells in chondrogenic medium seemed to form more continuous networks of cells and ECM in the capsule interior than did the same capsules cultured in growth medium. These results suggest that the cells differentiate in culture and produce ECM material, which shows the potential of the system to be used to regenerate cartilage.

478916 Perfusable Cells

Abstract:

A rapid vascular casting approach that uses polymer microparticle as a sacrificial template allows cells to survive in suspension until needed for transplantation. Blood vessel transplantation is the most convenient treatment for blood vessel defects. The defected area has recently been treated with autologous vessels or synthetic grafts, but the materials used should support cell adhesion, cell growth activity, cell differentiation, and maturation to the desired phenotype. To combat these challenges, we focus on engineering scaffold microparticles (MPs) to provide structural support to human umbilical-vein endothelial cells (HUVEC) while in suspension. The perfusable MP-embedded HUVEC use the U.S. food and drug administration (FDA)-approved biocompatible and biodegradable poly(lactic-co-glycolic acid) (PLGA) polymer as scaffolding. PLGA MPs of approximately 50 μm were produced using the solvent diffusion method. A bright field microscope and scanning electron microscope (SEM) were used to characterize the MP size and shape, respectively. The surface of MPs was conjugated with fibronectin, a glycoprotein involved in wound healing, to enhance HUVEC adhesion. Bioconjugation was accomplished by coupling primary amines of fibronectin with carboxyl groups of PLGA, forming amide bonds. The conjugation reaction was validated and quantified by using the Bradford assay. Fibronectin per milligram of PLGA MPs was calculated as 58.7 μg. HUVEC were seeded onto fibronectin-conjugated particles *via* gentle mixing and incubation, and the resulting cell-particle system was imaged with bright field, phase, and fluorescence microscopy over a period of one to two weeks for visualization of seeding efficacy and cell behavior. HUVEC exhibit promising cell adhesion on fibronectin-coated PLGA particles. Our results indicate perfusable cells can be cultured in suspension *in vitro*. More studies are under consideration on quantification of the cell population, molecular characterization, transport through blood vessel mimicking channels *in vitro*, and *in vivo* animal models. The perfusable cell therapies represent a possible solution in vascular tissue engineering.

478943 The Effect of Calcium Chloride in Modified CHO Cell Growth for Antibody Expression

Abstract:

The antibody is a protein molecule produced from CHO-cells as a secreted molecule and it is further purified to ensure the product quality attribute for the patients. During manufacturing of the protein, it has been shown that cell growth is a key factor to obtain the appropriate protein concentration of the molecule. Some cell growth discrepancies have been associated to specific nutrients within the CHO-cells culture media. Previous data indicates that the concentration of $CaCl_2$ in the media may impact negatively CHO-cells growth. Therefore, this investigation was focused in measuring the growth of CHO-cells using cell culture media spiked with different $CaCl_2$ concentrations. The experiment was performed in shake flasks at constant temperature, CO_2 injection and base media content. The viable cell density, viability, pH, pO_2, pCO_2 and metabolites were measured for day 0 and day 3. After performed the experiment it was observed that the concentration $CaCl_2$ affect cell in terms of the viable cell density (VCD). Statistical analysis shows that for 0M of $CaCl_2$ less cell growth occurred, but for concentrations near 1M the cell growth was optimum.

479535 Geometric Tools for Controlling Platelet Adhesion

Abstract:

Surfaces are ubiquitous in nature and technology. While natural surfaces excel in self-renewal and preventing bio-fouling, synthetic materials placed in contact with complex fluids like blood quickly foul. We present a novel biophysics inspired strategy for synthetic surface renewal using actuating surface topography. Using experiments, simulation, and theory, we study a general mechanism by which an adhered adlayer detaches from a surface undergoing wrinkling. We calculate a critical surface curvature beyond which surface renewal occurs. The scaling law is well validated using finite element simulations and physical experiments. While the de-adhesion model is deterministic, a statistical analogue is developed to study the effect of topography actuation in dynamically growing fouling adlayers such as platelet aggregates from blood contact. Using a system of actuating cylindrical grafts with varying wrinkle wavelengths placed in contact with whole blood, we show a dramatic over 90% decrease in platelet surface adhesion with topographic actuation. Moreover, the predictions of our statistical model are in strong agreement with the blood experiment data. Our cylinder data complement a second set of platelet adhesion data performed on PDMS surfaces and general mechanism is validated using macroscopic silicone patch experiments. The composite cylinders have been tuned for optimal actuation with arterial pulse pressure variation to function as a novel bypass graft.

479590 Investigation of the HP1-SUV39H1 Interaction in S. Cerevisiae

Abstract:

The eukaryotic genome exists as chromatin, a complex structure of DNA wrapped around octamers of histone proteins called nucleosomes. Biochemical modifications to N-terminal histone tails, including methylation and acetylation, help control eukaryotic gene expression, thereby regulating a variety of cellular processes. One important modification, methylation of histone H3 lysine 9 (H3K9), contributes to the generation of silent heterochromatin and has recently been implicated in the silencing of tumor suppressor genes and cancer. In humans, SUV39H1 is an H3K9 methyltransferase, and the protein HP1 interacts with SUV39H1 to help establish large, stable domains of methylated H3K9. The drug chaetocin is thought to inhibit the methyltransferase activity of SUV39H1and could be a potential anti-cancer drug. A better understanding of the molecular interactions between HP1, SUV39H1, and chaetocin is necessary to both illuminate chromatin-based transcriptional regulation and stimulate the downstream development of anti-cancer drugs that target cancer-causing chromatin modifications. Additionally, a better understanding of chromatin-based transcriptional control could result in novel gene regulation techniques for cellular engineering. Previous studies in natural cellular systems as well as in vitro approaches provide a wealth of knowledge about this heterochromatin system but cannot avoid either pleiotropic effects within cells or non-physiological conditions in vitro. Therefore, we have developed a synthetic two-hybrid system in Saccharomyces cerevisiae, a eukaryotic species that naturally lacks H3K9 methylation and its associated heterochromatin system. Our system has verified the direct interaction between specific domains of HP1 and SUV39H1 in living cells. Interestingly, we found that chaetocin, originally thought to inhibit the H3K9 methyltransferase activity of human SUV39H1, actually inhibits the interaction between SUV39H1 and HP1 and may therefore function through this mechanism to ablate heterochromatin formation. Our work has provided insights into the molecular interactions of an important heterochromatin system and provided new insights into a class of potential anti-cancer compound. Furthermore, the approach we have developed could be harnessed as a unique gene regulatory system for cellular engineering and biotechnology applications in yeast.

479592 Optimizing Nanoparticle Surface Characteristics to Minimize in Vitro Cytotoxicity

Abstract:

The study of nanoparticles is a tremendous field for discovery involving the efficient and effective delivery of therapeutics to specific target sites. Not only can modified nanoparticles minimize the affects of systemic blood circulatory delivery, but they also induce longer lifetimes of the therapeutic once delivered *in vivo*. Silica nanoparticles (SNPs) have been chosen for this project due to inherent biocompatibility, thermal stability, low toxicity and high range of functionalization. Specifically for drug delivery, the target size of these nanoparticles is between 50nm to 100nm. Through the use of the Stöber method, the target range of particles has been produced by adjusting the concentrations of various reagents within the reaction. Since surface characteristics have proven in literature to significantly affect performance of these nanoparticles within the human body, polymer coatings, such as poly(ethylene glycol) (PEG), have been examined as coatings to improve the half-life upon injection of the therapeutic. Although this approach is highly utilized within present research, the mechanism is not well understood, and there is even less knowledge concerning which PEG lengths are appropriate and optimal. Through this research, progress will be made in determining the optimal PEG length appropriate for SNPs to efficaciously minimize *in vitro* cytotoxicity, leading to future applications within the human body. Cellular uptake and viability studies are being conducted with particles covered in various lengths of PEG. These experiments are expected to produce a standard comparison on the effects of differently sized PEG. Within these experiments, Chinese Hamster Ovary cells are used due to their widespread use in cytotoxicity and uptake studies throughout literature. Comparisons between particles taken up by cells and those simply associated with cell surfaces will be made using confocal microscopy and flow cytometry. In order to efficaciously determine effects PEG length may have on particle uptake and/or toxicity, significant characterization will be conducted, including silica and PEG concentration determination.

479632 Studying Breast Cancer Metastasis Via the Effect of Breast Cancer Cells on Endothelial Cells

Abstract:

Metastasis, the phenomenon by which cancer cells spread from a primary tumor to distant organs, is greatly influenced by endothelial cells, which provide tumors with the oxygen and nutrients that they need to survive and spread. In this study, we analyzed cell proliferation rates of human umbilical vein endothelial cells (HUVECs), lymphatic endothelial cells (LECs), and microvascular endothelial cells (MECs) when co-cultured (CC) or cultured with conditioned media (CM) from MCF7, SKBR3, and MDAMB231 breast cancer cells. Cell proliferation was determined using a fluorescence assay and was quantified on days 1 and 4 of the experiment. We found that in LEC-CM and HUVEC-CM experiments, the more metastatic cancer cell lines increased cell proliferation as compared to control. In HUVEC-CC, MEC-CC and MEC-CM, all cell lines showed an inhibitory effect on endothelial cell proliferation, and in LEC-CC, MDAMB231 cells showed decreased cell proliferation as compared to other cancer cell lines. Very different trends in cell proliferation were seen for conditioned media versus co-culture conditions and, therefore, we can conclude that conditioned media and co-culture conditions do affect endothelial cells in different ways. In co-culture, growth factors secreted by the cancer cells can cause paracrine factor secretion by the endothelial cells, which could then stimulate a cyclic crosstalk between them that is not seen in conditioned media experiments. The results of this study will be used to correlate endothelial cell proliferation, among other phenotypes associated with cancer progression, with the associated metabolic profiles of the endothelial cells. The broader aim of this project is to link specific metabolic pathways and related regulation of specific factors to the physical behavior of the cells. This could lead to more effective clinical treatments for breast cancer through the targeting of endothelial cell functions critical for cancer cell survival.

479665 Optimization of Biomaterial Microenvironment for Motor Neuron Tissue Engineering

Abstract:

Injury and disease in the nervous system results in loss of sensory and motor neurons. Tissue engineering strategies have great potential to develop more effective therapeutics by combining appropriate biological cues with responsive biomaterial architectures. We have recently shown that more in vivo-like sensory neuron responses are found in 3D culture vs 2D flat substrates. Unfortunately, primary motor neurons are harder to isolate and culture compared to sensory neurons. Moreover, PC-12 cells are a well-characterized sensory neuron-like cell line and provide an alternative to primary sensory neurons. An analogous motor neuron-like cell line, NSC-34, is available but has not yet been characterized for tissue engineering applications. Thus, our goal was to systematically characterize NSC-34 cells and optimize biomaterial microenvironment and culture conditions for tissue engineering applications. For 2D studies, we cultured NSC-34 cells on coverslips coated with laminin or type I collagen in DMEM with fetal bovine serum (FBS; 3%, 10%), or N2 supplement. In our 3D studies, cells were cultured in gels composed of type I collagen with or without laminin or type IV collagen. We found that laminin substrates were associated with a higher percent of cells expressing neurites and that the neurites were significantly longer than for type I collagen substrates, with greatest results at 3% serum. Medium with N2 was associated with lower cell proliferation vs serum-containing medium regardless of substrate. Although proliferation results showed low percent proliferation for N2 medium, no significant difference was found between type I collagen and laminin. Preliminary results in 3D gels indicate that type IV collagen provides a favorable environment for neurite expression. Ongoing work focuses on rigorously characterizing neuronal morphologies and neurite lengths in 3D cultures.

479683 Synergistic IL6/8 Paracrine Signaling Pathway Infers New Strategy to Prevent Cancer Metastasis through Inhibition of Cell Migration

Abstract:

Following uncontrolled proliferation, primary tumor cells invade neighboring tissues by acquiring phenotypic changes that enhance migration. Tumor cell proliferation and migration have been previously shown to be mutually exclusive of each other. This study demonstrates that as tumor cells proliferate, past a threshold cell density, migration is enhanced in metastatic human sarcoma and carcinoma cells - but not in normal cells or non-metastatic cancer cells-, through a synergistic paracrine signaling via Interleukin 6 (IL-6) and Interleukin 8 (IL-8). Transcriptome sequencing reveals that the functional phenotype of high-density cells emerged due cell proliferation resembles that of low-density cells treated with a combination of IL-6 and IL-8. Mechanistic study with cell models indicates that the productive, paracrine IL-6/8 signaling pathway is only activated in 3D collagen matrices via Janus kinase 2 (JAK2), signal transducer and activator of transcription 3 (STAT3), Wiskott-Aldrich syndrome protein family member 3 (WASF3), and the Arp2/3 complex which induces enhanced protrusion activity and significantly increases cell migration. Combined inhibition of IL-6 and IL-8 receptors significantly decreases the expression of Arp2/3 in a mouse xenograft model and consequently reduces metastasis. This study reveals previously unobserved direct coupling between proliferation and migration through a synergistic paracrine IL-6/8 signaling pathway, which could be a therapeutically target for preventing or treating metastatic cancers.

479689 Electrospinning Stimuli-Responsive Fibers at the Nanoscale As Functional Drug Delivery Mats

Abstract:

Skin tissue engineering and drug delivery are very relevant fields of research in the health industry. Due to certain responses of the immune system, there has yet to be a development of proper skin tissue substitutes. A challenge that arises with this area of research is replicating a three-dimensional scaffold that has the same function and structures as the natural extracellular matrix. The scaffolds provide an alternative to the use of allografts and autografts and can be also used for bone repair.[2] A technique that has gained attention within this field of research in the last couple of decades is electrospinning. This new technology allows scaffolds to be created through electrospun nanoscale fibers with a large range of diameters, which produces a porous structure with a high surface area.[1] In addition to research of skin tissue engineering, drug delivery proves to be another promising application of electrospun fibers. The high loading capacity and localized Ð but time released Ð delivery of a large amount of therapies, along with the cost effective processing technique of electrospinning, make scaffold-facilitated drug delivery highly attractive. Dependent on whether degradable or non-degradable materials are use, the drug release can occur via diffusion or scaffold degradation.[3] The delivery rate at which the drug is released can also be altered, allowing for a larger range of drugs able to be delivered such as antibiotics, anticancer drugs, proteins and DNA.[3]

The objective of this research is to create biodegradable mats with tunable characteristics such as fiber diameter and surface area. The functional drug delivery mats enable disease-tailored therapies with targeted delivery to reduce side effects in patients. Using a large electric potential to draw fibers from a solution flowing at a specific rate, the solution reaches a grounded target several inches away. The biodegradable controlled drug delivery polymer used was poly(lactic acid-*co*-glycolic acid) (PLGA). PLGA solutions of 11 Ð 14 wt.% were prepared by dissolving the block copolymer in a solvent mixture containing tetrahydrofuran (THF) and dimethylformamide (DMF) at a 3:1 volumetric

ratio. They were then electrospun at distances of 7 and 18 cm and rates ranging from 0.8 to 4 mL/h at a voltage of 15 kV. The nanoscale fibers will be used as drug delivery mats and the kinetics of the peptideÕs release-time will be tuned to occur in the range of one hour to a week. In addition, solution rheology was performed on each PLGA solution to measure viscosity, which is directly correlated to the fiber diameter of the electrospun mats. Observing the impact of solvent on fiber spinning and fiber diameter brings about many positive results in developing fully characterized and well-understood fibrous mats for drug delivery.

Work Cited

[1]Ru, C.; Wang, F.; Pang, M.; et. al; Suspended, shrinkage-free, electrospun PLGA nanofibrous scaffold for skin tissue engineering. *American Chemical Society* **2015**, 10872-10977.

[2]Gentile, P.; Chiono, V.; Camagnola, I.; Hatton, P., An overview of PLGA-based biomaterials for bone tissue engineering. *International Journal of Molecular Sciences* **2014**, 3640-3659.

[3]Sill, T.; Recum, H., Electrospinning: Applications in drug delivery and tissue engineering. *Biomaterials* **2008,** *29.*

479694 Xylitol: Low-Calorie Sweetener

Abstract:

Xylitol: low-calorie sweetener

D. Gonzalez

Department of chemical and environmental engineering, Universidad Nacional de Colombia, Bogotá.

The xylitol is a polyol that contains 5 carbons, it can be obtained by reducing D-xylose sugar. It is classified as an alternative sweetener, because their properties are very similar to sucrose since it has a sweet taste, so you can get to replace fully contributing to food, main field in which it is used, the conditions. They are to be obtained from sucrose and reducing negative effects on the health of consumers.

This is very stable due to the absence of reducing groups, such as aldehydes and ketones, which are present in the reducing sugars. Xylitol has a heat load of 2.4 kcal / g and a relative sweetness equal to that of sucrose, so is the polyalcohol with greater sweetening power.

The World Health Organization recommends eating different types of sugars to reduce the risk of noncommunicable diseases. This is concerned with the prevention of weight gain up to unhealthy levels. It also establishes that dental diseases, whose occurrence is associated with the consumption of traditional sugars, non-communicable diseases are most prevalent in the world. (WHO, 2015)

Below is a comparison table where some of the representative characteristics of sucrose, sweetener comprehensive global consumption, and xylitol are discussed with respect to their effects on human health is presented.

saccharose	Xylitol
negatively affects health, causing: • Diabetes • Damage in the oral cavity • Obesity	safe sweetener for human consumption, in appropriate amounts, can promote the health of both children and adults.
It is metabolized by the bacteria Streptococcus pneumoniae and Streptococcus mutans, which cause ear infections and dental cavities respectively.	As a polyalcohol is not metabolized by certain species of bacteria such as those mentioned for sucrose.
There is no proven a laxative effect	It has been found that excessive consumption produces a laxative effect.
Not suitable for people who suffer from diabetes mellitus or are deficient in glucose 6-phosphate dehydrogenase	It is suitable for people who suffer from diabetes mellitus or are deficient in glucose 6-phosphate dehydrogenase

Xylitol is used as a sweetener in the food industry and also pharmacological level for the formulation of dental products as dentifrice, mouthwashes and whiteners. In the cosmetics industry it is used as humectant and hygroscopic due to its viscosity as emulsifier.

Industry	Application	Function
food	sweetner	generate sweetness
	Emulsifier / Thickener	form emulsions
Cosmetics	microbiological stabilizer	Inhibits bacterial growth
	wetting	moisturizing effect
Pharmacy	anticariogenicity	inhibits the formation of caries
	laxative	osmotic lubricant

479696 Optimizing Expression of Therapeutic Antigen-Binding Fragments (Fab) in E. coli

Abstract:

Antigen-binding fragments (Fab), proteins containing the functional region of antibodies, have been proposed as potential therapeutics for neurodegenerative diseases such as Alzheimerâ€™s. However, overexpression of Fabs has proven to be difficult. Previously, it has been demonstrated that certain folding engineering strategies have significantly improved functional expression of secreted proteins. This project tests the effect that the secretion pathway and the presence of trigger factor have on the overall expression of Fabs. Genes encoding the light and heavy chains of the C6T Fab were cloned into the pBLN-200 vector, each preceded by the same signal sequence, using Gibson Assembly. This was done with both the stII signal sequence (directing through the Sec-dependent pathway) and the DsbA signal sequence (directing through the SRP-dependent pathway). Both new plasmids were transformed into two different *E. coli* strains: BW25113, containing trigger factor, labeled tig^+, and KTD101, a BW25113 derivative lacking trigger factor, labeled Î"*tig*. These four systems were expressed at 37Â°C, and cell growth was closely monitored. Interestingly, the Î"*tig*-DsbA system, which was expected to have the highest efficiency of protein expression, produced the poorest cell growth, while the traditional tig^+-stII system demonstrated the most robust cell growth following induction. Future work will examine the levels of functional Fab overexpression and Fab export to the periplasm, which will help to explain the reason for this decrease in cell growth.

479976 Surface Characterization of WLBU2 and GP34 Chemically Immobilized on to Tcvs Modified SiO2

Abstract:

GP34 is a bacteria-binding tail-spike protein from bacteriophage that binds to the lipid A region of lipopolysaccharides (LPS) in bacterial membranes. N_3-Modified GP34, synthesized by Genetic Code Expansion, was covalently tethered through a copper catalyzed azide-alkyne addition to a brush layer of PBD-PEG diblocks with a terminal alkyne group. The PBD-PEG diblocks were bound to a TCVS modified surface through gamma irradiation. The biocompatibility of the GP34 was tested against WLBU2, a cationic amphiphilic peptide, which was previously being tested for endotoxin binding as well. WLBU2 was attached onto a brush layer through a conventional NHS tether. FITC Fibrinogen was then added to both brush layers to test for the repulsion of the protein, which would provide an early indication of the biocompatibility of the proteins. The absorbance of the supernatant and sample were measured using spectrophotometry at 490 nm to test for the fluorescence in solution and quantify the fibrinogen repelled. Results indicated that the brush layer with the click tethered GP34 performed better in repelling fibrinogen than its WLBU2 counterpart.

479989 Distillation Process Understanding from Analytical Monitoring and Computational Modeling

Abstract:

Aspen is a modeling tool used across multiple industries to simulate batch and continuous processes. When utilized during pharmaceutical small molecule development, an Aspen model can reduce the amount of laboratory testing needed before scale up by accurately predicting process parameters. For this particular study, an Aspen model was developed to understand solvent composition during a batch distillation process. Specifically, following the synthesis of an intermediate grade API (active pharmaceutical ingredient), a vacuum distillation is eventually performed to solvent swap from the extraction solvent (tert-butyl methyl ether, TBME) to the crystallization solvent (acetic acid). The goal for this work was to understand the molar ratio of TBME to acetic acid as the distillation progressed to determine the end point of the distillation without offline solvent composition analysis. In addition to Aspen modeling, real time composition information was obtained by monitoring the composition of the vapor condensation on the tip of an IR probe inserted into the reaction vessel. This study will illustrate how analytical monitoring by online IR and offline NMR has been combined with computational modeling in ASPEN Batch Modeler to generate process knowledge for the vacuum distillation, a key unit operation that needs to reduce the TBME:acetic acid molar ratio below a specified target to meet process quality goals. Knowledge around the distillation profile generated as a result of this study contributes to general process knowledge and can potentially eliminate the need to confirm the distillation endpoint via offline solvent composition analysis.

479993 Role of Gradient Microenvironments in Supporting Termite Gut Community Diversity

Abstract:

Termites can metabolize woody materials for energy, but no conventional biofuel production process has been able to emulate this functionality with comparable efficiency. The lower termite *Reticulitermes flavipes* is able to utilize 74-99% of the cellulose and 65-87% of the hemicellulose from lignocellulose with the help of microbes living symbiotically in the gut. If these microbes can be cultured *in vitro*, then they may be used in an engineered bioprocess to convert woody waste materials into useful products including biofuels and chemical manufacturing feedstocks. The hindgut of the lower termite *Reticulitermes flavipes* is a diverse microenvironment containing both aerobic and anaerobic regions in close physical proximity. Here we describe development of microhabitat engineering systems to measure survival of constituents of the termite gut microbiome as a function of oxygen gradient conditions. Replicate microhabitats were designed with either varying or constant oxygen conditions within the range from 0-10% equilibrium oxygen saturation in water. Various oxygen saturations were prepared using a microfluidic gradient mixer, and gradients were created by diffusion of oxygen from different concentration feeds through PDMS. Devices were loaded with a mixed microbial community collected from the hindgut of *Reticulitermes flavipes,* then viable protists were enumerated over for 24 hours. Oxygen concentrations were monitored optically using an oxygen-quenching fluorescent reporting molecule embedded in a polystyrene film. This work provides a basis for future developments in gradient bioengineering using the *Reticulitermes flavipes* microbiome, and may contribute to the sustainable production of biofuels from waste biomass.

480092 Ionic Liquids for Controlled-Release Drug Delivery: Synthesis and Solubility

Abstract:

Controlled-release delivery systems slowly release drug into the body over a long period of time, offering a possible treatment option for patients with chronic conditions who need constant medication. Currently, controlled-release drugs are usually implemented through slowly dissolving or degrading polymeric particles/pills or through osmotic devices, which are limited in which parts of the body they can be effective. While there are drawbacks, controlled-release over long periods of time is especially effective in that compliance and bioavailability is vastly increased over traditional formulations that need to be administered frequently through pills, patches, or eye drops. For example, ocular drug delivery is an excellent area into which controlled release can be applied, as many eye conditions require patients to apply multiple doses of eye drops a day, whereas controlled-release would require drug application on a much less frequent basis. Because of this higher reliability of patient care, controlled-release drug delivery is a highly researched field in pharmaceuticals.

Room temperature ionic liquids (RTILs) are one of the key focuses in this research; RTILs are liquid formulations of two ionic species and are characterized by high viscosity, high relative concentration of drug in formulation, and very low solubility in water. Their low solubility means that RTILs are a prime candidate for delivery of a drug via controlled-release, as they will naturally release small amounts of drug over time without the need for an osmotic or polymeric device. The high viscosity and drug concentration that RTILs exhibit also allow them to remain in a specific area and provide an effective drug concentration over an extended period of time.

The objective of our experiments was to synthesize RTILs using Diclofenac, an anti-inflammatory drug, and to characterize their solubility in aqueous environments over time. Two RTILs were synthesized with two different cationic surfactants (RTIL A and RTIL B) combined with the anionic Diclofenac. The RTILs were synthesized by combining equimolar amounts of Diclofenac and the complementary surfactant in methanol, and the RTIL products were measured to contain a 1:1 ratio of anion to cation. The solubility of the RTILs was evaluated by placing a saturated amount of RTIL in an aqueous solution and measuring concentration of the ionic liquid in solution over time. Concentration measurements were taken using reverse-phase HPLC, and the RTILs were found to reach a saturation concentration of about 1.5mM after around 10 hours. After this solubility curve was determined, the release profile of the drug from the RTIL in aqueous solution was also characterized, and found that both RTILs reached about 8% cumulative release after an 8-day period in the *in vitro* apparatus used in this study. Based on the solubility and release profiles, RTILs are a prime candidate for further research into their application as controlled-release formulations for ocular drug delivery, and studies should be done on their *in vivo* pharmacokinetics and pharmacodynamics.

480222 Theory-Driven Synthesis of Surface-Modified Nanoparticles for Biochemical Applications

Abstract:

Alzheimer's disease (AD) is the sixth-leading cause of death in the United States and accounts for over 60% of cases of dementia. Aggregation of amyloid-β protein (Aβ) in the brain parenchyma and cerebrovasculature has been indicated as a major etiological factor in AD, in accordance with the "amyloid hypothesis." Protein monomers assemble to first yield particularly toxic soluble species—oligomers and protofibrils—then larger insoluble fibrils, which deposit as the senile plaques characteristic of Alzheimer's pathology. Current AD treatment attempts only to alleviate symptoms of memory loss and problems with logic and reasoning; much of Alzheimer's research at the present seeks to obtain a greater understanding of the Aβ aggregation process, and evaluate compounds that inhibit it.

In previous research, gold nanoparticles (AuNPs) with weakly-associated electrostatic layers have demonstrated potent modulation of amyloid aggregation, at stoichiometric ratios as low as 1:2,000,000 nanoparticles per protein. AuNPs featuring a poly(acrylic acid) (PAA) outer coating, in particular, demonstrated inhibition of Aβ aggregation in environments with low ionic strength, while strongly promoting aggregation in aqueous salt solutions. This unique dual behavior makes these structures especially attractive as therapeutic candidates. From the effects of salt presence, we hypothesize that poly(acrylic acid)-coated gold nanoparticles alter Aβ aggregation by destabilizing the local solution environment. However, questionable stability of the electrostatic layers and complexity in modeling approaches suggest a simpler design of PAA-AuNPs with greater structural and experimental integrity.

This project describes the modeling and synthesis of neutral gold nanoparticles functionalized with end-tethered poly(acrylic acid) chains. Citrate-stabilized nanoparticles were synthesized using a modified Turkevich method procedure, and sized using TEM imaging. Acrylic acid polymers were synthesized via RAFT polymerization from a specialized thiol-containing chain transfer agent (CTA). The identity of the PAA sample was confirmed through NMR and FTIR spectroscopy, and average polymer size was measured via size exclusion chromatography. Polymers were then grafted to the nanoparticle surface, replacing citrate ions in a ligand exchange process. Functionalization to yield PAA-AuNPs was confirmed via UV-Vis spectroscopy.

The previous hypothesis is confirmed in part by study of such nanoparticle systems with a self-consistent mean field theory model for molecular dynamics. This approach performs a statistical thermodynamic analysis of a randomly-generated protein ensemble on a gold surface, as well as all species in the surrounding solution. Calculations of equilibria reveal that the polyelectrolyte brush "pulls" species in toward the surface to relieve negative charge density, affecting qualities such as pH of the local solution environment. However, in the presence of salt species, small sodium ions easily screen the effects of charge, resulting in a totally different solution profile.

Ultimately, the PAA-AuNPs synthesized in this project will be directly introduced into an aggregation mixture with Aβ monomer. Formation of larger structures will be measured via fluorescence spectroscopy. The combined theory and experiment will yield a better understanding of amyloid aggregation mechanics, and a process by which to optimize the design of an aggregation inhibitor.

480235 Biocompatibility and Biodegradation of Sensors Enabled By Zinc Oxide Thin Films

Abstract:

Biocompatibility and biodegradation of sensors enabled by zinc oxide thin films

Pooja Nanjannavar, Tanin Izadi, Vitaliy Avrutin, Nastassja Lewinski

Implantable biosensors for long term monitoring of disease biomarkers could improve survival rates and outcomes by giving individuals the ability to monitor their own health and take preemptive action against disease. However, the human body is known to reject implanted objects through foreign body reactions, biofouling, and inflammation. The objective of this study was to characterize the biodegradation rate of zinc oxide thin film coated biosensors to determine whether implanted biosensors could lead to potential cytotoxicity in human cells. We have measured the dissolution rate of zinc oxide thin films after 24 hours of incubation in cell culture media. We also tested the cytotoxicity of these extracts as well as the cytotoxicity resulting from direct contact of the coated sensor with cells. Zinc oxide concentrations in the media were measured using ICP-OES. Cytotoxicity was determined by measuring changes in metabolic activity using the tetrazolium-based MTS assay and changes in membrane integrity using the lactate dehydrogenase assay after 24 hours of exposure. Results from the tested sensor extracts show that there is little cytotoxicity (> 80% cell viability) associated with zinc oxide dissolution. Direct growth of cells on sensors preincubated with media also resulted in no observable difference in growth compared to the negative control. These findings demonstrate in vitro biocompatibility of zinc oxide thin film coated sensors for short term exposure. Future work will assess longer term biocompatibility and biostability.

480273 Development of a "Mix and Read" Assay for Target Detection Using Split-Luciferase Reconstitution

Abstract:

Current detection systems used to study protein-protein interactions utilize an elaborate array of chemical or biological labeling of the target for study, and thus have limitations in applications and analysis. To address the shortcomings of the typically used detection systems, such as an ELISA, the goal of this project was to develop a â€œmix-and-readâ€ assay where a target protein would be detected in a single step, without the need for washes. The system is based on a luciferase complementation assay using binders derived from the Sso7d protein scaffold fused to the components of NanoLuc split-luciferase. Luciferase is a bioluminescent enzyme that is found in many marine organism and arthropods, which scientists have dissected into two non-active components. These components, when brought into close proximity via protein-protein interactions, recombine to form the active luciferase molecule, which catalyzes a light-emitting reaction. For this protein complementation assay, Sso7d-derived binders that recognize a specific target will be fused to the N and C terminus of the split luciferase respectively. When binders are in solution with the target and the two binders bind the target, the luciferase molecule will reconstitute to its active form and with the addition of the reagent, a luminescent signal will be emitted. As a proof of concept, split components of NanoLuciferase, isolated from the deep-sea shrimp *Oplophorus gracilirostirs,* were fused to two lysozyme Sso7d binders. These preliminary findings show that lysozyme could be detected by this approach even after 3 minutes after the addition of the detection reagent, and the luminescence response increased with lysozyme concentration and incubation time. After validating the system, the next steps involve the development of a systematic approach to isolate Sso7d binders for other targets that could be used with the split luciferase model.

479837 Influence of Electrostatic Interactions on Particle Tracking for Microrheology Analysis

Abstract:

Influence of Electrostatic Interactions on Particle Tracking for Microrheology Analysis

Nur Hamideh[a], Yi-Syuan Guo[a], Leslie M. Shor[a,b]

[a] Department of Chemical and Biomolecular Engineering, University of Connecticut, Storrs, CT USA

[b] Center for Environmental Sciences and Engineering, University of Connecticut, Storrs, CT USA

Particle-tracking microrheology can be used to measure solution viscosity in-situ with high spatial resolution. In this technique, thermally-induced Brownian motions of fluorescent nano- or micro-sized spherical particles are tracked using an optical microscope. Viscoelastic drag on the spherical particles reduces the mean squared displacement of particles based on the Stokes-Einstein relation. This technique is increasingly used in biological applications using a range of different viscoelastic solutions, tracer particles, and chamber materials and dimensions. Here we evaluate accuracy of particle-tracking microrheology measurements as a function of charge interactions between tracer particles and the sample chamber. Two kinds of fluorescent tracers were used: 0.5 Î¼m diameter negatively-charged carboxylated polystyrene (PS) beads and 0.2 Î¼m diameter positively-charged aminated PS beads. Experiments were performed in an experimental assembly comprised of two glass microscope slides sealed with silicone grease. Both glass surfaces were pre-treated by applying 25 Î¼L aliquots of four different types of biological solutions (tryptic soy broth, lysogeny broth, yeast extract, and bovine serum albumin), spread with a sterile swab, and allowed to dry. Particle trajectories of 50-80 particles were quantified in 40% glycerol-water solution and 58 Î¼g/mL

exopolysaccharide (EPS) solution using a Zeiss Axio Observer.Z1 inverted microscope and 40x objective, then analyzed using Fiji ImageJ and MatLab. Deviations between viscosity measured using particle tracking microrheometry and viscosity measured using traditional bulk-scale viscometry will be presented. These results will give insights into modulation of charge interactions by biological solution surface treatment in particle tracking microrheology.

480346 Development of Nanoclusters Using Tannic Acid-Coated Ultra-Small Iron Oxide Nanoparticles for Simultaneous MRI Imaging and Drug Delivery

Abstract:

Iron oxide nanoparticles are generally believed safe and effective magnetic resonance imaging (MRI) contrast agents and are in clinical use for iron deficiency treatment. We have shown that ultra-small nanoparticles (USNPs, < 4 nm) are great positive (T_1) contrast agents, the clinically preferred type. However, these extremely small nanoparticles experience fast renal clearance, limiting their use in therapy tracking. Here, we are forming nanoclusters by crosslinking individual small nanoparticles in order to increase the size, which subsequently improves their circulation times. In this presentation, we will discuss the method of synthesizing the ultra-small nanoparticles, surface functionalization of the nanoparticles, and the formation of the nanoclusters. The nanoclusters will be characterized by size, size distribution, and relaxivity. Bovine serum albumin (BSA) and tannic acid interactions will facilitate the formation nanoclusters and the procedure can be modified to tailor the size of the nanoclusters. The clusters were successfully formed within the desired size range with T_1 character.

480420 Evaluation of Curcumin-Loaded Nanoliposomes for the Treatment of Age-Related Macular Degeneration

Abstract:

Age-related macular degeneration (AMD), the most common cause of vision loss for people age 50 and over, is a disease characterized by the buildup of oxidative stress in the back of the eye. Current remedies are limited to intravitreal injections that only target the more severe 'wet' form; the common 'dry' form has no pharmaceutical solution. Curcumin, a natural antioxidant, has shown potential in combating inflammatory diseases like AMD; however, the molecule also demonstrates poor bioavailability. This study synthesizes, characterizes, and tests a liposomal complex encapsulating curcumin to improve its delivery to dry AMD-affected retinal cells via noninvasive methods. This project will study the complex's potential to protect retinal pigment cells from damage and its ability to permeate the layers of the eye in order to prevent or delay the onset of dry AMD.

480437 Passive Mechanical Properties of Polymer-Coated Cells for High Speed Sorting Using Antigen Specific Lysis (ASL)

Abstract:

Introduction

ASL is a new cell sorting technique that sorts cells by coating the target cells with a temporary cross-linked polymer, and then exposing the entire cell population to harsh conditions to lyse the uncoated cells. Purity of ASL with surfactant based lysis exceeds 95% while the batch process takes 1h. We seek to eliminate damaging surfactant residues by using osmotic lysis. The objective of this study is to relate the mechanical properties of polymer coatings to the prevention of osmotic lysis.

Methods and Materials

Micropipette aspiration technique was used to test the cortical tension of both uncoated Jurkat cells and polymer-coated Jurkat cells. The data was analyzed using Laplaceâ€™s Law and the calculated cortical tension is compared. The hypothesis is that by providing a greater tensile reinforcement to the cells, high modulus coating yield more live cells than low modulus coating during a hypotonic treatment.

Cells were prepared at a concentration of ~500,000 cells/mL, they were washed twice with PBS and then suspended in 1mM EDTA and 0.5% FBS. Then the cell solution was placed in a chamber on the microscope stage, a glass micropipette prefilled with PBS is used to aspirate cells. The pressure is created by two water reservoirs connected to the pipette and is measured by a pressure transducer.

Jurkat cells were polymer coated by first labeling the cells with photoinitiator (eosin) through standard immunolabeling against CD45 with eosin probes. Cells were then suspended in a pH 7.4 phosphate buffered solution of PEG diacrylate (Mn 575, 3500, or 8000), triethanol amine, and vinyl pyrollidinone. Polymerization commenced with irradiation (20 mW/cm2, 10 minutes, 530 nm LED lamp). Upon rinsing, the polymer coated cells were suspended in the same solution as the uncoated cells and were then tested.

Tension values were calibrated using the neutrophilsâ€™ cortical tension is ~15Â±3 pN/Âµm, which falls in the lower end of literature values.

Results/Discussion

The cortical tension of Jurkat cells was tested and was found to be ~20±4 pN/µm. Upon coating with a PEG (Mn 575) diacrylate coating, the cortical tension increased to 700±49 pN/µm; upon coating with a PEG (Mn 3500) diacrylate coating, the cortical tension increased to 2917±274 pN/µm. Cells protected with both PEG diacrylate coatings resisted swelling in pure deionized water and retained viability.

Conclusion

The polymer coating prevents cells from osmotic lysis by providing tensile reinforcement. By finding the correlation between tensile reinforcement and polymer type and thickness, we can potentially control the mechanical properties of the cells by changing the coating and thus potentially be able to sort cells more effectively, which has great potential for future clinical applications.

479898 Biocompatibility of 3D Printer Material to Bacterial Cultures

Abstract:

The pace of innovation in microfluidics and other lab-on-a-chip fields has been limited by the long, labor-intensive production cycles and the expensive laboratory equipment needed for traditional microfabrication techniques. Production of microfluidic devices via 3D printing can drastically reduce the cost and prototype production time for microfluidic masters over photolithography. More rapid iteration of cell handling and culturing devices accelerate the pace of biotechnology innovation as well, but only if cells can be effectively held and cultured in 3D printer materials. Here we examine the effects of surface treatment on bacterial growth inside custom 3D-printed well plates produced from Clear2 Resin in a Formlabs Form2 stereolithography (SLA) printer. Treatment methods focused on removal of free crosslinking mediators, and included shining light of UV spot cure laser, agitation in water for 24 hours, and a combination of both treatments. *Escherichia coli* and *Pseudomonas fluorescens* bacterial cultures were inoculated into treated, untreated, and polycarbonate control well plate chambers, and growth kinetics were collected using a plate reader. Experiments revealed that the untreated proprietary material decreases the growth rate of the bacteria strains to varying degrees. To quantify the biocompatibility of the material for the different strains, we calculated the maximum growth parameter and the doubling time of the bacteria in each of the well conditions. This data indicated that biocompatibility with the less vigorous *E. coli* strain can be improved by utilizing novel surface treatments on the 3D-printed material. We anticipate that our well design and biocompatibility assay could be used as a starting point for labs interested in measuring the biocompatibility of a specific organism or organisms before adopting 3D-printed geometries for different biological research applications.

479925 Optimization of the Î±-Synuclein Antibody Stability through the Isoelectric Point for Parkinson's Disease Applications

Abstract:

Recent studies show that the aggregated and misfolded proteins within the brain directly and indirectly kill neurons, primarily through fibril formation one being α-Synuclein (αS). Studies of brains afflicted by Parkinson's disease have shown that protein fibrils form years before the disease actually is present. Utilizing *E. coli* bacterial expression, an antibody fragment was isolated that targets anti-α-Synuclein. This fraction is called a single chain variable fragment (scFv). It is easily produced, but does not have long-term stability. The primary goal of this investigation will be to determine new dialysis conditions that will ensure better stability of the scFv. Applying the ideal dialysis method could potentially create a more stable protein. Identifying an ideal isoelectric point through dialysis will benefit the overall stability of the structure.

This process starts with expressing the αS and then filtering the supernatant with the tangential flow filter (TFF). Afterward, the scFv is purified with ion metal affinity chromatography, utilizing nickel bead. The scFv bound to the nickel bead column, and is then released with the imidazole. Subsequently, the purified fraction is dialyzed in a variety stable storage buffers, including PBS, Tris, and HEPES. For each solution, the pH is adjusted to determine the effect on the stability of the purified fragments. The primary goal was to determine the isoelectric point, the settings where the proteins possess the equal amount of positive and negative charges. The electrostatic attraction contained in the protein can cause aggregation and produce precipitates, therefore rendering the engineered antigens useless for further study, therefore process conditions were accounted for throughout the study. An SDS-PAGE gel and western blot are used for confirmation of the identity of the protein, and a BCA test is applied determine protein concentration. Then, enzyme immunoassays (ELISA) are utilized to determine the stability and functionality of the scFv purified from the various buffers. The accumulated results of the ELISA were then compared to determine the most ideal dialysis conditions for anti-αS. This outcome became a standard procedure for future production of this particular protein.

After the series of analytical tests to confirm identity, functionality, and concentration, the results of the project concluded that phosphate-buffered saline at pH 7.4 provided the highest yield of scFv. Further studies must focus on the improvement of the stability of α-Synuclein through by identifying and improving laboratory protocol.

480442 Polymer Nanoparticles Co-Coated with Antibodies and Anti-Phagocytic Moieties Improve Specific Interactions and Uptake By Target Versus Phagocytic Cells

Abstract:

Nanocarriers (NCs) have great potential for drug delivery to improve therapeutic effects. However, rapid clearance of drug NCs by the mononuclear phagocyte system (MPS) remains a major challenge. Phagocytic cells in the liver and spleen engulf and eliminate NCs out of the bloodstream, reducing their circulation time and therapeutic potential. Poly(ethylene glycol) (PEG) functionalization on NCs lowers phagocytosis and lengthens circulation. However, PEG can reduce specific NC interactions with target cells and render generation of anti-PEG antibodies. In nature, circulating erythrocytes avoid phagocytosis by expressing on their surface protein CD47. Similarly, NC coated with CD47 have reduced phagocytosis in the liver and spleen. Yet, these studies had not targeted NCs and whether CD47 coating alters binding and endocytosis of antibody-targeted NCs is unknown. To evaluate this we used fluorescent polystyrene or PLGA NCs targeted to ICAM-1, a receptor overexpressed on many diseased cells. Anti-ICAM/CD47 NCs were compared to control preparations (anti-ICAM NCs, IgG NCs, IgG/CD47 NCs) to determine phagocytosis by macrophages versus specific endocytosis by endothelial cells using fluorescent microscopy. Results showed that, compared to anti-ICAM and IgG NCs, presence of CD47 significantly reduced phagocytosis by macrophages (~50% reduction vs. absence of CD47). However, anti-ICAM/CD47 NCs had similar targeting to endothelial cells than control anti-ICAM NCs (~87% of control by 30 min). Anti-ICAM/CD47 NCs were endocytosed by endothelial cells similar to control anti-ICAM NCs (~99% of control at 5h). Furthermore, anti-ICAM/CD47 NCs used the same CAM-mediated endocytosis pathway as anti-ICAM NCs (~40% inhibition by amiloride versus no significant inhibition of clathrin pathway by MDC or caveolar route by filipin). Therefore, our results show that coating NCs with CD47 and a targeting antibody helps reduce macrophage phagocytosis while providing effective binding and uptake to target cells. This strategy can have a significant impact on improving the efficacy of targeted drug delivery systems.

480443 Alternative ELISA Using an RNA Aptamer Against Calf Intestinal Alkaline Phosphatase

Abstract:

Alkaline Phosphatase (AP) is extensively used in enzyme-linked immunosorbent assays (ELISA), most requiring expensive and time-intensive reagents to generate (such as AP-conjugated antibodies). In addition, these antibodies may have reduced immunoreactivity when conjugated with AP. We seek to address these problems by developing an alternative AP ligand in the form of an aptamer (nucleic acid binding species) against Calf Intestinal AP (CIAP). This CIAP-aptamer will lead to the development of an alternative reagent that is readily amendable to conjugation, such as to a primary antibody or perhaps another aptamer.

Anti-CIAP aptamers were isolated using Systematic Evolution of Ligands by Exponential Enrichment (SELEX). After nine rounds of selection, we found a reoccurring 14-mer RNA motif that presented itself in a loop structure for the majority of RNA sequences. Aptamers 4-3, 4-9, 3-6, and minimized variant 3.1 bound to CIAP with Kd values of 5 nM, 9.4 nM, 10.8 nM, and 6.8 nM, respectively.

We designed a nucleic acid construct consisting of our CIAP aptamer along with a linker and extension, which hybridizes with an anti-PDGF-AB DNA aptamer. Preliminary results show an expected linear correlation between PDGF-AB concentration and absorbance at 405 nm, though absorbance readings are low. Future experiments include increasing the sensitivity and limit of detection of the assay, and improving the versatility of our reagent by swapping the PDGF-AB aptamer with a 2â€™-fluoropyrimidine anti-FC aptamer.

480447 Enhancing Ethanol Production with RNA-Engineering of Zymomonas Mobilis

Abstract:

A key challenge facing the biofuel industry is the antimicrobial nature of ethanol and other pretreatment chemicals to the microbes that catalyze the conversion of biomass to ethanol. *Zymomonas mobilis* has been recognized as a useful bacterium for this catalysis because it naturally produces and tolerates ethanol in high quantities. Small regulatory RNAs (sRNAs) in *Z. mobilis* that change expression under ethanol stress indicate the possibility to engineer increased ethanol production and tolerance. In this study, we evaluate which sRNAs most positively affect ethanol production by overexpression or deletion from the genome. This way, we can engineer a more robust strain of *Z. mobilis* that is apt for ethanol production by combining different overexpressions and deletions of these RNA regulators. In doing so, we serve the biofuel production community seeking to develop the most efficient microorganisms to enable chemical production in a cost-effective and sustainable way. By increasing ethanol yields directly and allowing bacteria to live longer in their own product, we demonstrate that multiple sRNAs can be manipulated to enhance ethanol tolerance and ultimately that sRNAs are powerful tools for metabolic engineering in this organism.

480449 Cross-Linking Amyloid Forming Proteins for Improved Understanding of Early Aggregation

Abstract:

Student: Gram Booth

Mentor: Dr. Christa Hestekin

Institution: University of Arkansas

Title of Project: Cross-linking amyloid forming proteins for improved understanding of early aggregation

Project Summary:

Misfolding of human amylin protein leads to aggregation, or interactions of proteins into a group, which is involved in a variety of diseases including Alzheimer's and type II diabetes. The around five million Americans that have developed Alzheimer's disease continues to increase with the aging population. In the U.S. alone, over 29 million people had been diagnosed with diabetes by 2014, most of which representing type II. In order to properly treat these diseases, understanding of the underlying mechanism for protein aggregation has been an important topic of study. Our project focusses on the difficult to analyze early stages of aggregation. Photo-induced cross-linking of unmodified proteins (PICUP) was used to study these early, unstable stages of protein aggregation by allowing us to take "snapshots" of the protein aggregation as it progresses. PICUP allows researchers to collect data about the smaller, more toxic stages that are still soluble and therefore more maneuverable in the body. PICUP allows these early protein aggregates to be analyzed for characteristics including: size distribution, relative abundance of the different sizes, and hydrophobicity. Results from this project may allow for the creation of better models to represent aggregation in diseases such as type II diabetes and Alzheimer's. The improved understanding of the aggregation mechanism could also lead to treatments for the reduction or inhibition of aggregation.

480462 Engineering of a Stable Myosin Neck Fragment for Understanding the Consumption of Energy in Resting Muscle

Abstract:

Myosin is an important protein component of muscle cells which works in conjunction with actin to allow muscles to relax and contract. Myosin consists of three regions, head, neck and tail. There is a relation between the conformation of the neck and relaxed, active, or super-relaxed states of myosin. Super-relaxed myosin has reduced energy consumption. The neck fragment we plan to study contains two molecules of the regulatory light chain (RLC) and two fragments of the heavy chain (HCF), which forms a coiled coil. The engineered fragment was to maintain the three-dimensional structure observed in the full length protein. In the full length protein one molecule of RLC binds one molecule of HCF. Two of these HCF/RLC complexes dimerize *via* a coiled coil formed by HCF chains, altogether resulting in a functional tetramer complex. We subcloned RLC and introduced two point mutations into the HCF amino acid sequence that increased its ability to form a coiled coil. Engineered RLC and HCF were co-expressed and co-purified on a Ni-NTA resin using the histidine tag located on HCF. This allowed for a 1:1 stoichiometry between the RLC and HCF in the purified RLC/HCF protein complex. Using mass-spectrometry, we have confirmed the presence of both RLC and HCF in the purified protein complex. Circular dichroism (CD) spectra of the purified protein complex indicated the coiled coil formation. Using the glutaraldehyde cross-linking reaction we have also shown that a tetramer complex was formed. Therefore, by engineering these fragments we were able to reconstruct a stable myosin neck fragment, small in size and amenable for further studies.

480518 PLGA Encapsulation of Apocynin

Abstract:

Drug delivery to the brain is a challenging task due to the blood brain barrier. PLGA is a biopolymer commonly used as a biodegradable drug delivery carrier. Study of the degradation of PLGA encapsulated drug is useful for drug applications. The drug Apocynin is used to treat Parkinsons. Apocynin is an anti-inflammatory drug used to treat colitis. The size and zeta potential of the encapsulated drug is of importance since passing blood brain barrier and mimicking human body conditions in essential for the drug encapsulated particle to operate correctly. In this experiment different concentrations of PLGA and APO are tested for size, zeta and concentration degradation. The particles are engineered by adjusting the concentration ratio of APO to PLGA. The optimal particle size for particles are 100 to 200 nm. The optimal formulation of particles is sought from the size, zeta, and concentration degradation.

480548 Optimizing the Quality and Throughput of Electrospun Poly(Glycerol Sebacate) Vascular Grafts

Abstract:

Electrospun poly(glycerol sebacate) (PGS) is used to fabricate vascular grafts. We optimized the quality and process of fabrication of our PGS grafts by decreasing time electrospinning and residual carrier polymer after curing. We accomplished our aims by improving the wash methods and implementing multiple-needle electrospinning. We showed that a more aggressive wash with a lower mass graft to solvent volume reduced the amount of PVA in the graft from the previous methods of washing of 9.7±2.5% to 5.8±0.47% by weight. Two needles were successfully spun at the same time, and essentially doubled the rate at which PGS/PVA solution was electrospun on to a mandrel when paired with a needle array. Fiber diameter and alignment were monitored with ImageJ. The graft will be implanted into a sheep model to test its regenerative potential.

480549 Mechanical Properties of Spherical Alginate Hydrogels

Abstract:

Hydrogel beads can be used as a method of drug delivery for humans. More specifically, the beads can contain bacteriophage that act as an antibiotic when released to help cure various medical issues. Studying the mechanical properties of these beads can help determine which type of bead is best for drug delivery, since typically digestion in the stomach lasts about 5 hours, then food remains in the intestines for up to 48 hours. Ideally, the bacteriophage would be released between the 6 and 48 hour time points after ingestion to avoid damage to the bacteriophage. The objective of this research is to study the behavior of calcium chloride and barium chloride beads cross-linked for different amounts of time while in lambda buffer solution. From the data collected, the best bead option(s) for bacteriophage release can be determined.

480550 Reducing Fibrous Encapsulation of Implanted Devices Using Biomaterials

Abstract:

Fibrous encapsulation, commonly referred to as scar formation, can cause failure in many biomedical implants, such as in the artificial pancreas for type I diabetes treatment. Scar tissue prevents wastes and nutrients from being exchanged in the implant, ultimately rendering the treatment painful or ineffective. For example, one potential therapeutic for type I diabetes involves using alginate to protect insulin-producing cells. This artificial pancreas provides diabetics with their much needed insulin on-demand. If scar tissue forms around this artificial pancreas, the device loses the ability to deliver insulin and becomes useless. Our goal is to reduce the fibrous encapsulation of implants using poly-l-arginine-based biomaterials and improve the understanding of how fibroblasts involved in scar formation respond to biomaterials. We show that our materials can reduce fibrous capsule formation by inhibiting the activation of latent transforming growth factor Î² (L-TGF-Î²). Active TGF-Î²Â leads to capsule formation by activating fibroblasts to differentiate into myofibroblasts, which contract collagen. We also discuss the viability of our biomaterials and presence of VEGF in vitro. This research will not only improve the success of the artificial pancreas, but other biomaterials.

480527 An Improved Method to Edit Undomesticated Lactobacilli Strains Using Crispr-Cas9 Systems

Abstract:

The rise in new genetic tools have given scientists powerful ways to modify and study prokaryotes. However, many of these methods require a high transformation efficiency of the bacterial strains in order to work, which significantly limits use of these methods across species. Here, we developed a novel method utilizing CRISPR-Cas9 and repair templates to genetically edit low transformation efficient strain of abundant *Lactobacillus Plantarum*. By using CRISPR-Cas9 in these strains, we show that single point mutations are almost impossible to achieve without using a double stranded DNA repair template. We plan to use this method to further probe and study the diverse lactobacilli genus. This method offers a powerful means of editing lactobacilli and should be translatable to other strains of difficult to work with bacteria.

480534 Multivalent Cation-Triggered Folding of Hydrogel Sheets

Abstract:

Self-folding gels have been created for purposes such as drug delivery, biosensing, and tissue scaffolding. These gels typically exhibit short-side folding, i.e. folding occurs about an axis parallel to the short side of the rectangle. The only examples of long-side folding in gels involve applying a constraint, such as clamping the gel to introduce a bias. In this project, we report the fabrication of a self-folding hydrogel sheet that transforms from a flat sheet to a tube, via long-side folding, and characterize this folding behavior. We believe that long-side folding occurs due to the nonhomogenous stresses as crosslinking of the gel occurs to initiate folding. The bilayer consists of two components, one of which interacts with multivalent cations such that the swelling ratio is affected to create differential swelling. We explore the behavior of folding in response to various multivalent cation solutions and the concentration of a biopolymer. These parameters allow us to tune the folding axis and tightness of the curled structure. Long-side folding opens up more possibilities in 3-D structures that bilayer gels are able to fold into. The stimulus of cations is also important for biomedical applications, especially if the gel can be optimized to respond to physiological concentrations.

480567 Effect of Rock Inhibition on Mass Transport in Mesenchymal Stem Cell Aggregates Undergoing Chondrogenesis

Abstract:

The objective of the research was to manipulate biomolecular signaling inside differentiating human mesenchymal stem cells (hMSCs) to improve properties of tissue engineered cartilage. Enhanced transport conditions formed by manipulating the biophysical mechanisms activated during hMSC aggregation upon chondrogenic differentiation were investigated. Manipulation was achieved through Rho/ROCK signaling, a key biophysical signaling pathway that interferes with actin-myosin mediated contractile force generation and subsequent aggregation (Kim & Sheetz, 2012). The effect of the Rho/ROCK inhibition has not been investigated yet in hMSCs, which highlights the importance of this study.

Chondrogenesis is the process by which chondrocytes are formed from progenitor cells. Currently, small aggregates of scaffold-free cartilage constructs can be formed by subjecting hMSCs to culture medium containing growth factors such as TGF-Î²1. Within 24 hours after chondrogenesis initiation, the cells aggregate spontaneously to form a spherically-shaped pellet. Aggregates have been used primarily as an investigative tool and not as a therapeutic tool due to their inferior properties compared to the native tissue.

While properties of hMSCs undergoing Rho/Rock inhibition were not previously investigated, hMSCs have been observed to form a â€œloosely packed pelletâ€. A more loosely packed pellet of hMSCs can be linked to greater transport properties allowing for signaling molecules to permeate aggregates and therefore improving conditions for cartilage tissue culture. While pellets being formed during the first few days will not have the structure necessary for transport properties to be observed, they could be fully observed after seven days. As the extracellular matrix accumulates during chondrogenesis, the transport properties will be inhibited, which occurs beginning day 7. Rho/Rock inhibition was able to significantly enhance the transport properties due a less compact pellet.Â The central hypothesis

is that modulation of transport properties via ROCK inhibition during chondrogenesis will lead to improved cartilage tissue. The resulting construct due to the Rho/ROCK signaling will lead to larger pellets, allowing the aggregates to be used as a more practical tool for clinical applications.

The goal of this study was focused on the effects of the inhibition of Rho/ROCK and its effect on hMSCs towards chondrogenesis and enhancing the process. This was evidenced by the increased mass transport of tracer dextran molecules of different sizes. Fully formed ROCK inhibited pellets were noted to have improved transport properties than those without the inhibition. Biochemical property data supports this above conclusion.

480612 Effect of Nodal Status Compared to Tumor Response after Neoadjuvant Radiation on Outcomes for Patients with Esophageal Cancer

Abstract:

Introduction: The Dutch CROSS trial established the role of neoadjuvant chemoradiotherapy for improved survival in the treatment of patients with advanced esophageal adenocarcinomas. Both surgery and radiation therapy represent means of loco-regional control, and esophageal cancer is disease that is associated primarily with systemic failure. The goal of the current study is assess the clinico-pathological features of patients with resected esophageal cancer undergoing neoadjuvant radiation therapy on long-term survival.

Methods: All patients at a single institution who underwent esophagectomy for esophageal adenocarcinoma after neoadjuvant radiation therapy from January 2004 to December 2014 were included. Patients were divided into short-term survivors (STS) who died of disease-specific causes within 1 year of resection and long-term survivors (LTS) who lived in excess of 5 years after surgery. Peri-operative deaths were excluded. Logistic regression analysis was performed to identify factors predicting LTS.

Results: Of the 334 patients who underwent esophagectomy for adenocarcinoma after neoadjuvant radiation, there were 74 (22.2%) STS and 28 (8.4%) LTS. STS were older (median age 65.5 vs. 56.6 years, P=0.003) and were more likely to have lymph node metastases prior to radiation (93.2% vs. 39.3%, P<0.0001) than LTS. LTS had a trend towards having a complete pathologic response to radiation (46.4% vs. 27.0%, P=0.10), undergoing an open operation (92.9% vs. 77.0%, P=0.09), and having a shorter post-operative length of stay (median 8.5 vs. 11.5 days, P=0.10). Univariable and multivariable analyses are shown in the Table. Only the absence of lymph node metastases on the initial evaluation correlated with LTS (P=0.04), whereas pathologic complete response to neoadjuvant radiation did not (P=0.18).

Conclusion: Response to neoadjuvant radiation therapy in patients with esophageal cancer does not correlate to survival. Presence of pre-treatment lymph node metastases is likely a marker of systemic disease and is the only variable that predicts outcome with patients with esophageal adenocarcinoma. Clinical response to radiation therapy should not dictate decision for esophagectomy.

480618 Foaming Agents for Drug Delivery to Treat Inflammatory Bowel Disease

Abstract:

About 1.6 million Americans suffer from Inflammatory Bowel Disease (IBD), a condition that includes Crohnâ€™s Disease and Ulcerative Colitis. Treatment of this condition focuses on reducing the inflammation caused by IBD either through anti-inflammatory drugs or immunosuppressants . Many times it is preferable to deliver these drugs topically,Â and foams provide several benefits over the common liquid enemas. A number of foam products are already on the market to treat colitis including foams containing Mesalazine and Budesonide . Our experiment focused on evaluating non-traditional foaming agents and formulating an optimal foaming solution for lower gastrointestinal tract drug delivery.

480686 Interleaflet Coupling Effects in Supported Lipid Membranes

Abstract:

Cellular membranes contain amphiphilic lipid molecules that self-assemble into a bilayer structure. In biological membrane systems, there has been considerable interest in phase-separated domains termed lipid rafts. These domains play an important role in cellular functions as organized centers for signaling molecules such as membrane proteins and influence membrane fluidity. Fundamental knowledge of a lipid domainâ€™s biophysical properties in model systems informs a greater understanding of their functions and properties in living systems. In this research project, we are investigating the properties of phospholipid bilayers by using substrate-supported lipid bilayers (SLBs) as model systems. In these model systems, domains are almost exclusively observed as being coupled across the bilayer (i.e. in phase-registry), suggesting that interleaflet interactions play a notable role in domain formation. To investigate the strength of interleaflet coupling in SLBs, we aim to develop an electrophoresis-based method to measure the coupling between the leaflets. Electrophoresis is conventionally used in biology to separate macromolecules based on size and electrostatic charge by application of an external electric field. In contrast, we attempt to use applied electric fields to induce directed, lateral diffusion of charged lipids and lipid domains. By monitoring the electrophoretic response of domains, we seek to characterize the relative strength of coupling interactions in different SLB systems. In the future, we plan to investigate the differences in characteristics of a membrane deposited on a fluid-phase and gel-phase monolayer in comparison to one deposited on a covalently grafted monolayer, comparing a dynamically responsive phospholipid system with a mimetic covalently grafted system. Additionally, we would like to explore the implications of depositing phase separated monolayers on inner leaflets that are not normally prone to domain formation. By observing microstructural changes in phase separated monolayers before and after deposition, we intend to gather information regarding the role of coupling in sustaining preexisting domains.

480577 Lactic Acid Fermentation of Sporolactobacillus in Glucose and Biomass-Derived Sugars

Abstract:

Lactic acid has a wide array of applications in the food-processing and pharmaceuticals, and its derivative, polylactic acid (PLA) can be developed into an environmentally-friendly biodegradable thermoplastic. Industrial lactic acid production has focused on utilizing bacterial fermentation rather than synthesis due to high manufacturing cost, less sustainability, and the inability to form the D-lactic acid stereoisomer.This D-lactic acid stereoisomer is combined with L-lactic acid to form stereocomplex polylactic acid, which can lead to a more thermally resistive plastic. The D-lactic acid stereoisomer is desired to form a stereocomplex polylactic acid, which could lead to a more thermally resistive plastic. This study attempted to test the D-lactic acid fermentation of a *Sporolactobacillus* strain in glucose media and HTM, a sugar cane biomass feedstock, before scale-up. D-lactic acid titer was measured by using YSI and HPLC, and growth of the strain was monitored by measuring OD and pH. Bacteria in HTM was found had a 68% higher productivity while bacteria in a glucose environment had a 22% higher yield. These results will help determine the future scalability of this strain.

480582 Effects of PLGA Composition, Concentration and BSA Loading on Electrospun Nanofiber Geometry and Controlled Drug Release

Abstract:

Abstract: The purpose of this research was to examine how the composition and concentration of the co-polymer poly(lactic-co-glycolic acid) (PLGA) affects the geometry of electrospun nanofibrous matrices for the purposes of drug delivery. The affects of loading bovine serum albumin conjugated with fluorescein (FITC-BSA) into these matrices on fiber geometry were also examined. These fibers were electrospun with an 18-gauge syringe tip, at a collecting distance of 10 cm, at a voltage of 20 kV and at a flow rate of 1.0 ml/hr. These fibers were imaged and analyzed using SEM to determine the average fiber diameter and diameter variability. It was determined that 20 wt% PLGA (50:50) solutions yielded fibers with the most desirable characteristics. At a concentration 15 wt% PLGA for both compositions, beading was present and at a concentration of 25 wt% the average fiber diameter for both compositions increased significantly. The average fiber diameter of 20 wt% PLGA (50:50) fibers was determined to be 1.05±0.19 Î¼m. The average fiber diameter for 20 wt% PLGA (85:15) was determined to be 1.08Â±0.38 Î¼m. Because of the smaller average fiber diameter and standard deviations of the PLGA (50:50) fibers, this composition of PLGA was used for the purposes of controlled drug release.

PLGA (50:50) solutions of 15, 20 and 25 wt% were then prepared and 0.5 mg/ml of FITC-BSA was added to each solution. The solutions were then electrospun under the same conditions stated previously. It was determined that 20 wt% PLGA (50:50) fibers loaded with FITC-BSA yielded the best fiber characteristics. The average fiber diameter was determined to be 550.48Â±91.04 nm. To ensure that FITC-BSA was evenly dispersed throughout the fibers, confocal microscopy was used. The fluorescent molecules could be seen in the fibers electrospun with FITC-BSA but could not be seen in the fibers without FITC-BSA. The 20 wt% PLGA (50:50) fibers loaded with FITC-BSA were then submerged in a physiological buffer (PBS) to determine the feasibility of using these fibers as a drug deliver system. It was determined that these fibers could act as a time released drug delivery system as the concentration of

FITC-BSA in PBS increased over time.

480591 Rapid Uptake and Ubiquitination of Fluorescent Peptides into Mammalian Cells Using a β-Hairpin Sequence Motif

Abstract:

Fluorescent peptide-based reporters have emerged as a prominent method to quantify intracellular enzyme activity in cells. Normally rapidly degraded by intracellular enzymes, inclusion of a β-hairpin 'protectide' on the N-terminus of a peptide substrate increases reporter stability inside the cell. Effective transport of these peptide reporters into intact cells has presented several challenges to researchers. For example, one method of transport is microinjection with a needle, but this process can damage the cell. However, a cell penetrating peptide (CPP) sequence can overcome these limitations. This research presents a library of short peptide sequences that use a β-hairpin motif as both a protectide and a CPP. Members of this peptide library demonstrated the ability to rapidly penetrate the cell membrane of intact HeLa cells independent of incubation time and temperatures assessed by both fluorometry and fluorescent microscopy. β-hairpin permeability efficiency was compared to commercially available CPPs [e.g., TAT and Arg(9)] and a negative control peptide known to not penetrate intact cells. Two peptides named RWRWR [Ac-RWVRVpGO(FAM)WIRQ-NH_2] and OWRWR [Ac-OWVRVpGO(FAM)WIRQ-NH_2] were discovered to exhibit the highest permeability efficiency. Peptide stability was demonstrated using a degradation assay coupled with reverse phase high performance liquid chromatography (RP-HPLC). The commercial peptides , while superior in cellular uptake efficiency, displayed showed reduced cellular stability. Additional members of the peptide library were used to show that cellular uptake was not dependent on the secondary structure of the β-hairpins, but the lack of the secondary structure showed decreased cellular stability. Additionally, orthinine-rich members of the peptide library have been previously discovered to act as primary degrons, resulting in *in vitro* ubiquitination. Here, it was determined that these peptides could not only gain entry to intact HeLa cells, but could also be rapidly ubiquitinated. This combination of attributes offers significant promise for these β-hairpin protectides to serve as next generation biosensors for members of the ubiquitin

proteasome system.

480725 Anti-Angiogenic Properties of Biofunctionalized Silver Nanoparticles

Abstract:

Studies have shown that silver nanoparticles (AgNPs) elicit a cytotoxic effect. The degree of cytotoxicity is dependent on size, concentration, time of exposure and surface chemistry. AgNPs can also be engineered to selectively interact with specific cell types by surface functionalization with biomimetic targeting moieties. To this end, citrate stabilized AgNPs were functionalized with thiol-containing cyclic RGD peptides by ligand exchange to target $\alpha v\beta 3$-integrin receptors on the cell membrane of hemangioendothelioma (EOMA) cells. Confocal microscopy and PCR were used to confirm the expression of the $\alpha v\beta 3$-integrin receptors in EOMA cells. Cells were incubated with 20nm citrate stabilized or RGD-AgNPs at variable concentrations and exposure times in serum-free media. An MTS assay was used to quantify the number of surviving cells following exposure via absorbance spectrophotometry. Higher cytotoxic responses correlated with increased nanoparticle concentration and exposure times. RGD-peptide functionalization to AgNP surfaces was demonstrated by an immunoblotting assay; however, there was not significant enhancement in cytotoxicity relative to non-binding RAD control peptides. This could be due to the non-integrin dependent mechanisms of cellular internalization reported for the cyclic RGD peptide surfaces and future work will be aimed at promoting integrin specific internalization utilizing antibodies.

480739 Optimized Approach to Synthesize Ultra Small Gold Nanorods to Target Nanospaces

Abstract:

Optimized Approach to Synthesize Ultra Small Gold Nanorods to Target Nanospaces

Priya Prabakaran, Ming Cheng, Eno Ebong

Abstract

Nanoparticles have recently become very important in the biomedical field due to their ability to deliver therapeutics to difficult areas. These areas include nanospaces which are nano-scale voids that are difficult to access. Depending on their application, nanoparticles can be made of different types of materials. For drug delivery applications, the surface of the gold particles can be modified since gold is highly reactive with thiols which forms strong bonds with drugs.[1] Other advantages of gold nanoparticles include applications for imaging and photothermal therapy.[2] Gold is also useful due to its high biocompatibility. In addition, many different shapes can be synthesized including spheres and rods both of which can be used for different applications.[3] Spheres and rods are useful because their size allows access to nano-sized voids in tissue. Rods have been found to be better for drug delivery due to their high surface area. The higher surface area, allows for more interaction between the binding sites and the membrane receptors. The size and shape of nanoparticles can be controlled using specific experimental conditions. However, little is known about how exactly different parameters affect the dimensions of nanorods or how to create consistently small rods in order to target nanospaces such as those found in cells and the extracellular matrix. Filling this gap is our main focus. This experiment aims to find a repeatable approach to synthesize nanorods to be used to target nanospaces.
For this experiment, the seed growth method was used to synthesize gold nanorods. A seed solution consisted of 25mM $HAuCl_4$, 0.1M CTAB solutions, and cold $NaBH_4$. A growth solution was comprised of 0.5 mM $HAuCl_4$, ascorbic acid, 0.12 mM $AgNO_3$, and 0.1M CTAB solutions. The ascorbic acid concentrations that were used were 0.6 M, 1.0M, and 1.2 M. The seed solution was then added to the growth solution which was placed in a water bath. The temperatures of the water bath were tested at 30, 33, and 35°C and the growth time was tested at 1hr, 2hr, 3hr, and 4hr. The nanorods were extracted by centrifugation and imaged using Transmission Electron Microscopy (TEM). This experiment quantifies the length and width of the nanorods relative to the growth temperature, growth time, and ascorbic acid concentration.

At the reference condition with a growth temperature of 33°C, a growth time of 1 hr, and an ascorbic acid concentration of 0.1 M, particles were synthesized with dimensions around 25nm by 55nm. All methods described above synthesized dimensions that were less than 8nm by 25nm. It was also determined that 35°C and 1.2 M of ascorbic acid gave the smallest dimensions. Time is not a factor that affects dimensions.

Future studies will be done to incorporate drugs and imaging agents to these small nanorods.

References

(1) Huang, X., et al. *Advanced Materials, (2009)* (2) Burrows, N. D., et al. *Langmuir* (2016)

(3) Pérez-Juste, J., et al. *Coordination Chemistry Reviews* (2009)

480597 Rheology of Bovine Cervical Mucus

Abstract:

Cervical mucus is the viscoelastic fluid secreted by the cervix. Dr. Leo Han at OHSU is working to evaluate human cervical mucus samples according to their Insler Score, a method that ranks the samples based on their consistency, ferning, spinnbarkeit, cellularity and volume. Protocols for storing the samples are lacking, as little information is known concerning the limits of storing the mucus at various temperatures. Currently, no protocol by the World Health Organization (WHO) exists for storing cervical mucus. Thus, the goal of this project is to determine whether or not freezing or refrigeration has any rheological effects on the mucus samples. Bovine cervical mucus, easily accessible in large quantities, was selected as a model replacement for human mucus. Using a standard rotational rheometer, dynamic oscillatory-shear tests were completed to determine trends between the viscous and elastic moduli of various bovine cervical mucus samples as a function of storage time and temperature.

481092 Protein Nanocarrier for Targeted Intracellular Delivery of Functional Antibodies

Abstract:

The cell membrane remains a formidable barrier for antibody-based therapies, and efficient intracellular delivery of functional antibodies would enable modulation of intracellular signaling mechanisms and protein-protein interactions involved in various disorders. This study utilized protein engineering techniques to develop a novel protein nanocarrier that is capable of delivering functional antibodies to the intracellular environment. Each nanocarrier contains six SPAB antibody-binding domains, and is therefore capable of delivering up to six antibodies. The interaction between the SPAB domain of the nanocarrier and the heavy chain constant region of the antibody is noncovalent, thus allowing the nanocarrier to bind different functional antibodies with the same affinity. Three iRGD domains were integrated into the nanocarrier structure to allow for selective targeting of integrins, which are commonly overexpressed in cancer cells. We successfully expressed the protein monomers, assembled the functional nanocarrier, and investigated its antibody-delivery properties. Results of cellular uptake studies involving HeLa, MCF-7, as well as SK-BR-3 cancerous cell lines indicate significant internalization of nanoparticle following 24-hour incubation. Flow cytometry data also indicate substantial cellular uptake of antibody-loaded nanocarrier as compared to soluble antibody control. In addition to efficient cellular uptake, the highly biocompatible and modular nature of our nanocarrier makes it ideal for expanding the scope of antibody-based therapeutics to the intracellular environment.

480761 Development of a Stimulus Responsive, in-Situ Forming, Nanoparticle-Laden Hydrogel Formulation for Opthalmic Drug Delivery

Abstract:

Glaucoma is a chronic optic neuropathy that ranks as the leading cause of blindness worldwide, affecting as many as 80 million people. Inadequate or obstructed drainage of the aqueous humor through the trabecular mesh within the eye increases intraocular pressure, which subsequently damages the optic nerve situated at the base of the retina. Expectedly, most glaucoma medications exert therapeutic action by lowering the intraocular pressure, and nearly all of these are administered as eye drops. However, since corneal penetration efficiencies of <5% are typical for such dosage forms, which implies that the dosage of drug in these formulations is much higher than required, there is a clear, unmet need for novel transcorneal drug delivery vehicles that achieve higher drug penetration at significantly lower dosages. Such a vehicle would greatly facilitate the formulation of newer, more effective but less bioavailable anti-glaucoma drugs. To this end, we have developed a stimulus-responsive, nanoparticle-laden hydrogel vehicle for spatiotemporal and dosage-controlled release of poorly bioavailable drugs into the aqueous humor of the eye. Our hydrogel matrix is formulated as a composite of two FDA-approved, GRAS polymers that are biocompatible and highly mucoadhesive. Similarly, the physical attributes of the drug-loaded nanoparticles too have been synthetically optimized in order to achieve high transcorneal penetration. Experimental design aided the identification of hydrogel composition and nanoparticle content in the formulation for stimulus-responsive switching between temperature-dependent rheopexy (thickening) and thixotropy (thinning). The liquid formulation has a sol-gel transition temperature of 32Â°C, which is the temperature of the ocular surface, and its gelation characteristics permit the formulation to form a thin, uniform coating over the cornea through blinking of the eyelid. These properties facilitate its application as an eye drop immediately prior to the patientâ€™s bedtime, and also permit easy and scalable manufacturing, which greatly simplifies the path towards development of a regulated industrial-scale manufacturing process. Significantly, our formulation exhibits over a 300% increase over control formulations in whole-eye experiments using Bovine eyeballs, thereby paving the way for the introduction of novel anti-glaucoma products to the market.

480763 Development of a Microfluidic Device to Quantify Algal Chemotaxis

Abstract:

The directed migration of algae in response to external chemical cues, or chemotaxis, can play an important role in the development of harmful algal blooms (HABs). These blooms pose a significant concern to human health and the environment due to the decrease in oxygen levels and increase in toxin production. It is well established that many environmental factors influence the formation and development of a bloom; however the behavior of algae within the bloom has not been thoroughly studied. One area of interest is the diel vertical migration and recruitment of algae to a bloom, both of which are governed by chemotaxis. Due to the limitations of current technology and assay techniques, direct measurement of algal chemotaxis has not been completely studied. Microfluidic devices, which allow for the precise control of the cellular microenvironment have been used in several studies to quantify the chemotactic response of other organisms. However, to date, this technology has rarely been adapted to the study of algal chemotaxis. Here, we present the development and optimization of a â€˜flow-freeâ€™ microfluidic gradient generator for the study of algal chemotaxis. The device consists of an agarose hydrogel bottom layer and a PDMS top layer imprinted with three parallel microchannels. A chemical gradient is developed by flowing media containing a source compound (ammonia for algae) in an outer channel and a sink solution in the opposing outer channel, providing two constant-concentration boundary conditions. An important feature of this device is that the center is a flow free channel, which exposes cells to a stable, linear gradient in the absence of direct flow. As a proof-of-concept, we demonstrate the utility of the device by exposingÂ *Chlamydomonas reinhardtii*, a model unicellular green alga, to a linear nitrogen gradient. Preliminary findings indicate that both gradient intensity and steepness influence the tactic response of the cells. Additionally, the migratory behavior of C. reinhardtii was found to be strongly influenced by the presence of ambient light. The direct observation of algal chemotaxis has not been possible to date, thus the results presented here provide new insight into the dynamic migratory behavior of algae.

480764 Quantification of Alkaline Phosphatase Activity in Single Algal Cells Utilizing a Microfluidic Device

Abstract:

Quantification of alkaline phosphatase activity in single algal cells using a microfluidic device

Travis Dugas, Kelly Yates, David Englehardt, B. Seth Roberts, Sibel Bargu, and Adam T. Melvin

Microfluidic devices are commonly used to decrease the incubation required and reagent volume needed to perform certain biochemical assays. The goal of this project is to develop a microfluidic device to trap single algae cells and quantify the activity of the enzyme alkaline phosphatase (AP) in the green algal species *Chlamydomonas reinhardtii* in the presence and absence of organic phosphorous. This assay is capable of assessing the activity enzyme alkaline phosphatase (AP), which plays an important role in informing researchers of the available levels of organic phosphorous available in the water column. In fact, measuring AP activity provides a more accurate metric of phosphorous levels in the water that can specifically impact the initiation and propagation of harmful algal blooms (HABs). As part of this study three microfluidic devices were fabricated. The first device fabricated was capable of isolating single algae cells in a trapping array consisting of 230 semicircular wells. AP activity was assessed in cells using a commercially available stain (AP Live Stain). AP activity was quantified across a population of, ~70 cells per experiment at increasing concentrations of organic phosphorous. These results confirmed that AP activity is directly proportional to the phosphorous concentration in the cellular environment. This device was also capable of finding variations in AP activity across an entire population. However, one significant challenge encountered during experimentation was a low trapping efficiency in the 230 trap array. In an effort to increase single cell trapping efficiency, numerical simulations were performed using COMSOL to obtain a more comprehensive understanding of the fluid flow dynamics within the microfluidic device. From these computational results, two new microfluidic devices were designed consisting of 120 and 60 cell trappers. The hallmark of this device was the introduction of two parallel streams at different velocities to create a sheath flow and increase trapping efficiency. Preliminary experimentation determined a trapping efficiency of ~70-80% in these new designs. Ultimately, this work demonstrates the utility of incorporating an enzymatic assay into a microfluidic device, modeled through COMSOL, to not only decrease the reagent cost and incubation time, but also to increase the overall trapping efficiency of single cells to be evaluated in a much shorter time.

480786 Investigating the Effect of Hydrogel Thickness on Bacterial Adhesion

Abstract:

The rise of antibiotic resistant bacteria is a growing epidemic that necessitates the use of alternative antibiotic-free approaches to delay the onset of bacterial adhesion and subsequent infection. This is especially true for hospital settings and the healthcare industry. Healthcare-associated infections (HAIs) stemming from medical devices, like catheters, are preventable financial and physical burdens that too often lead to morbidity. In practice, the surface of these catheters are coated with thin, lubricious polymer hydrogels to passively prevent microbial adhesion. However, the prevention of bacterial adhesion using the poly(ethylene glycol) (PEG) hydrogels erodes over time, leading to significantly increased infection rates with prolonged catheterization. While many chemical alterations on this coating have been tested, their basic geometry and material properties have remained constant, namely their thickness and stiffness. It has previously been determined that key cellular processes, including adhesion and proliferation, are mechanically responsive for both mammalian and microbial species. However, the depth of this mechanical sensitivity has not been investigated. Therefore, the aim of this current study is to elucidate the effect hydrogel thickness has on bacterial attachment and subsequent growth. This was accomplished through a series of bacterial attachment and mechanical characterization experiments of hydrogels with stiffness of 40 – 6500 kPa and thicknesses of 10-150 μm. For the bacterial attachment experiments, gram-negative *Escherichia coli* and gram-positive *Staphylococcus aureus* bacteria were utilized due to the highly different mechanical sensitivity mechanisms the species exhibit. By also accomplishing *in situ* atomic force microscopy (AFM) characterization of hydrogel surface features for each relevant hydrogel, we were able to decouple the effect hydrogel thickness and hydrogel stiffness has on bacterial adhesion. Our findings indicate that bacteria "feel" through a hydrogel to an underlying support layer unless a sufficiently thick hydrogel is utilized. Determining bacteria's depth-dependent mechanical sensitivity mechanism holds potential to improve the design of next generation medical device coatings that can limit bacterial adhesion without the use of antibiotics.

477601 Kinetic Study of Catalytic Decarboxylation/Decarbonylation of Triglycerides to Fuel-like Hydrocarbons

Abstract:

The use of algae to capture CO_2 from coal-fired power plants constitutes an interesting solution to the necessary innovation in the field of carbon capture and utilization, particularly because algae can intermediate the conversion of these emissions into valuable fuels and chemicals. Although a renewable fuel, biodiesel (fatty acid methyl esters) possesses a high oxygen content which renders it less than ideal as a transportation fuel. Moreover, current methods to deoxygenate algae lipids, such as hydrotreating, are costly due to the requirement for high pressure hydrogen. Decarboxylation/decarbonylation represents an advantageous alternative because the process produces second generation fuels without the costly requirements of hydrotreating. This project aims to lay the groundwork for kinetic studies of the catalytic conversion of pure triolein (a model compound of the extracted lipid feed) in dodecane to optimize the catalytic process through which algae oil is converted to fuels. The nature of these studies requires the use of a system optimized to limit diffusion limitations and act with negligible transport resistances at steady-state. These properties are dependent upon reactor and particle size, catalyst dilution, and the nature of the reaction (heat effects, equilibrium conversion, vapor-liquid equilibrium). Establishing the kinetic parameters of this reaction, such reaction orders and activation energies, can aid in the identification of the kinetic pathways that afford high product yields.

477608 A Two-Step Approach to Demetallation of Algal Biocrude

Abstract:

Much of humanityâ€™s progress in the past century has relied on use of fossil fuels. Although these sources have provided undeniably cheap and convenient access to energy, their continued use has contributed to problems such as smog, acid raid, and the rising level of greenhouse gasses. Biofuels have emerged as a renewable and potentially carbon-neutral alternative to fossil fuels. Specifically, biofuels made from algae are advantageous because of their high lipid density, the ability to be grown on low-value land, and the fact that the sourcing doesn't interfere with food supply. Although algal biofuels have shown promise, the field is still relatively new. One ongoing problem that has stagnated adoption is the high metal content inherent in the sourcing of these fuels. Before algal biofuels can become commercialized, a cost-effective method for removing metals in the biocrude must be developed. This project addresses multiple methods for biocrude demetallation, and investigates which conditions provide the least metal contaminated products.

477623 High Temperature Measurements of the Reaction of Heptanone Isomers with Hydroxyl Radicals

Abstract:

Ketones are considered potential biofuel candidates and may be produced using lignocellulosic biomass. In the current study, the reaction rate coefficients of three heptanone isomers (4-heptanone, 2-methyl-3-hexanone, 2,4-dimethyl-3-pentanone) with hydroxyl (OH) radicals are investigated experimentally. During the combustion and oxidation of hydrocarbon fuels, the highly reactive OH radicals oxidize the fuel molecules representing the major fuel consumption pathway under combustion conditions. A shock tube facility at King Abdullah University of Science and Technology (KAUST) was used as a batch reactor and a ring-dye laser system was applied to create a narrow line-width laser beam at 307 nm to follow the concentration of OH radicals as the reaction progresses. Since OH radicals are unstable species, tert-butyl hydroperoxide (TBHP) is employed as an OH radical thermal precursor. A mixture of ~300 ppm of the ketone and ~15 ppm of TBHP is prepared in argon diluent, and the mixture is then introduced into the shock tube. Reaction rate coefficients are then measured under pseudo-first-order kinetics over a temperature range of 851-1328 K and pressures near 1.5 atm. Available detailed chemical kinetic model for 3-pentanone is upgraded with sub-mechanisms of TBHP and the three studied ketones. The modified chemical kinetic model is used in conjunction with the CHEMKIN-PRO commercial program to simulate OH profiles. Furthermore, a fitting procedure is used to obtain the best match between the simulation and experimental OH time-history profiles in order to extract the reaction rate coefficients of the following reactions: 4-heptanone + OH --> Products (R1) 2-methyl-3-hexanone + OH --> Products (R2) 2, 4-dimethyl-3-pentanone + OH --> Products (R3)

The Arrhenius forms of the obtained rate coefficients are:

$k_1 = 3.15\times10^{-10}\exp(-2949\ K/T)\ cm^3\ molecule^{-1}\ s^{-1}$,

$k_2 = 7.92\times10^{-11}\exp(-1552\ K/T)\ cm^3\ molecule^{-1}\ s^{-1}$,

$k_3 = 7.67\times10^{-11}\exp(-1883\ K/T)\ cm^3\ molecule^{-1}\ s^{-1}$.

Finally, the experimentally measured reaction rate coefficients are compared with a well-known rate estimation method, known as Structure-Activity Relationship (SAR).

477626 Bio-Briquette Production As Renewable Energy from Waste of Banana Peels and Adhesive of Molasses

Abstract:

The banana culture wastes can be used to produce briquettes as fuel for energy generation. Biomass is one of the sources of energy production with the greatest growth potential in the coming years and can be easily obtained from agricultural production, which generates large amounts of waste. The use of agricultural and agro-industrial wastes as biomass fuel for power generation is being increasingly studied and could be an alternative solution to the problems. These wastes can be availed as briquettes. Briquettes produced from this waste at low cost are an excellent source to produce cheap energy following environmentally correct way and they are, in many case, ideal for replacing fossils fuels in use today, with significant economic and environmental advantages. Among the advantages of using briquettes, it can be mentioned: reduction of thr deforestation, due to subtitution of the wood generally used; production of cheaper energy; reduction of the environmental impact caused by the large amount of wastes and its destination; and reuse of leftover materials.

According to Soffner (2001) on his journal "Pulp Production from Banana Stem", the bananas generates a significant amount of wastes, approximately 200 t/ha/year, since each plant produces one to five bunches of bananas. The most significant waste and more bulky are leaves, pseudostem, peel and stalk. A large amount of banana peel has been left as garbage after industrial processes. The raw material is plentiful and has low economic value. Therefore, the characteristics of banana-peel briquettes including banana peel properties were investigated. Aiming the reuse of wastes, in this work, banana peel was still used and characterized to produce briquettes as fuel in several applications.

477992 The Effect of Cyclic Molecules and Branched Alkanes on the Gelation of Petroleum-like Solutions

Abstract:

During the transportation of petroleum using subsea pipelines, wax deposits can form to block the pipelines at low temperatures. These gels can reduce oil production and are very costly to handle in the petroleum industry. Although numerous components are present in the wax deposits, their primary components are n-alkanes which have low solubility and simple structures.

However, other components in the petroleum are known to influence the properties of wax deposits. In this work, different combinations of n-alkanes were used to represent petroleum waxes using octacosane (C28), dotriacontane (C32) and hexatriacontane (C36). The aim of this work was to investigate the role of cyclic molecules and branched alkanes have on the n-alkane gelation. Phenanthrene (PAH) and squalane (SQ) was chosen to represent cyclic molecules and branched alkanes respectively.

It was found that Both PAH and SQ do not cocrystalize with C32 or C36. Significant amounts of PAH or squalane are needed to influence the gelation of the co-crystalizing n-alkane system. At high concentrations of PAH, the gelation become dependent on PAH. The presence of cyclic molecules or branched alkanes eliminates the dispersant effect of the shorter-chained n-alkane on the longer-chained n-alkane.

478048 Synthesis of Lithium Titanium Oxide-Based Composite As Anode Material for High Performance Lithium Ion Batteries

Abstract:

Rechargeable lithium ion batteries are regarded as reliable energy storage and supply systems for large scale applications due to their long cycle life, high gravimetric and volumetric energy densities, and low self-discharge rate. On this subject, spinel lithium titanium oxide $Li_4Ti_5O_{12}$ (LTO) is a commercially successful "zero-strain" Li-ion intercalation anode material, whose present large scale application is limited by its low theoretical capacity of 175 mAh g^{-1} at a high operation potential of 1.55 V vs Li+/Li. In the present study, we try to incorporate silicon (Si) nanoparticles into the LTO host matrix by using a high energy ball milling process under inert atmosphere, since Si has the highest known specific capacity of 4200 mAh g^{-1}, and boost the conductive properties of the electrode through addition of polyacrylonitrile (PAN) conductive polymer. As a result, a combination of LTO, Si and PAN will supplement each other and afford a composite anode materials with a superior Li-ion storage capacity, enhanced conductivity and excellent mechanical properties. For instance, LTO/Si/PAN composite with weight ratio of 25:25:50 showed initial capacity of 2150 mAh·g^{-1}, and after 30 cycles capacity was 1200 mAh·g^{-1} at a current density of 50 mA·g^{-1}, when cycled between 0 and 3 V (vs. Li^+/Li).

478318 Establishing a Co-Culture System for Clostridium Cellulovorans and Clostridium Aceticum for High Efficiency Biomass Transformation

Abstract:

C4 bio-refinery, such as butyrate and butanol production, is currently a research hotspot. However, undesirable waste products including H_2 and CO_2 result in lower product yield and a waste of energy. *Clostridium aceticum* has the ability to convert H_2 and CO_2 to acetate, which can be assimilated by other clostridia strains through the reversible acetate kinase and phosphotransacetylase pathway. This study setup a co-cultural system for *C. aceticum* and *C. cellulovorans*. Together they can convert the cellulosic biomass to butyrate or even butanol if a metabolic engineered strain is used, with considerably higher yields. Based on DSMZ 520 medium for *C. cellulovorans* and DSMZ 135 acetobacterium medium for *C. aceticum*, the practical pH ranges were evaluated for both strains. The pH 7.0-7.5 was suitable for both strains. By mixing these media with various ratios, an optimized formulation was defined for the co-cultural system. The evaluation of co-culture was conducted using glucose as substrate. The conversion rate of glucose to final products has increased by 5-10%.

478367 Removing Roadblocks to Make the Conversion of CO2 Emissions to Diesel and Jet Fuel Using Algae an Industrially Viable Process

Abstract:

Algae-based carbon capture represents one possible solution to the issue of rising greenhouse gas levels. In addition to capturing CO_2 emissions, this approach facilitates CO_2 conversion to algal biomass, the components of which â€“ including algal lipids â€“ can be monetized. However, the overall efficiency and economic constraints of extracting, purifying and upgrading algal lipids present challenges that must be overcome for algae oil to become a viable source of fuels and chemicals. The work described in this contribution includes the optimization of algal lipid extraction and purification processes designed to avoid the destruction of valuable algae components and the use of an improved catalytic approach to obtain drop-in hydrocarbon fuels from the purified algal lipids. Low-cost, scalable methods are employed, including the direct transesterification of wet algae biomass to extract algal lipids and the use of low-cost adsorbents to purify the crude algal lipids. Raw algae, as well as crude and purified algal lipids, were analyzed throughout in order to observe the evolution of the lipid profile and the elemental composition of the feed as it was processed. Similarly, the spent adsorbents and catalysts used during the purification and upgrading of the algal lipids were analyzed as a means to gain insights into the final destiny of metals and heteroatoms undesirable for fuel applications. Activated carbon was found to be an affective adsorbent for the elimination of impurities such as chlorophyll, phospholipids and sterols from the crude algae oil. Notably, comparison of the UV-Vis spectra of a chlorophyll standard, the crude algal lipids and the purified algal lipids showed a complete absence of chlorophyll in the purified sample. Moreover, analysis of the purified lipids and the product of their catalytic upgrading revealed the latter to be ideal for transportation fuel (diesel and jet-fuel) applications.

478427 Tailoring Electrolyte Composition to Improve the Activity and Energy Efficiency for the Electroreduction of CO_2 to CO

Abstract:

Atmospheric carbon dioxide (CO_2) concentrations have been on a constant rise in the past few decades. This rise in CO_2 levels has been correlated to an increase in the global temperature anomalies leading to adverse climate change effects such as rising sea levels and erratic weather patterns. As a result, developing sustainable chemical technologies that can utilize and eventually reduce the excess CO_2 emissions is one of the grand engineering challenges of the 21st century. Electroreduction of $CO_2$to value-added chemicals (such as carbon monoxide (CO), methanol, formic acid, and ethylene) using renewable energy could provide an interesting solution to this problem.

In this work, we will explore the effect of electrolyte composition on the electroreduction of CO_2 to CO using two different state of the art cathode catalysts (*i.e.*, silver nanoparticles and gold nanoparticles supported on polymer-wrapped multi-walled carbon nanotubes). Our results indicate that the use of alkaline media (high pH) improves the activity and lowers the overpotential requirements for the electroreduction of CO_2 to CO. Additionally, we will also report results regarding mixtures of deep eutectic solvents with inorganic salts and how that can be used to integrate the process of CO_2 capture with conversion.

478830 Heat Transfer in Rotary Drums Via Conduction, Convection, and Radiation

Abstract:

Granules are the second-most used substance after water [1] in many industries like the food, pharmaceutical, ceramics, and construction industries. Granular processing consumes roughly ten percent of the worldÕs energy [2]. Granules, however, donÕt behave like conventional solids, liquids, or gases, which makes them difficult to model. Thus, there is a great need for research in optimization of these processes to save energy and resources. Rotary drums are often used for granular processing including milling, heating, cooling, mixing, granulation, and more. However, heat transfer in rotary drums can be inefficient, leading to wasted energy. Valuable resources can be saved by developing more accurate models for the heat transfer in rotary drum applications to granular processing.

This research focuses on understanding the heat transfer mechanisms in the granular bed, inside a rotary drum, via conduction, convection, and radiation. Experiments will be performed using silica beads in a stainless steel drum to investigate the effect of drum rotation rate, granular fill level, and inlet air flow rate on heat transfer between the drum wall and granular bed. Silica is a common catalyst and most industrial drums are stainless steel, so this research represents common industrial applications.

Granular processes can be modeled using discrete element method (DEM) simulations by inputting particle physical properties such as density, viscosity, and thermal conductivity. The graduate student leading this research has modeled conduction in a rotary drum and is working on the models for convection and radiation. These models must be validated by experimental data in order to prove their accuracy, which will be the focus of this specific study.

Once the simulations and validations have been accomplished, experimental results will be compared to simulated results in order to examine the validity of the simulations and the models can be applied to quantify all three modes of heat transfer and study which modes dominate under varying operating conditions. The results can be used to optimize rotary drum efficiency in larger-scale industrial applications.

References

[1] Andreotti et al., Granular Media: Between Fluid and Solid, 2013.

[2] Duran, Sands, Powders, and Grains: An Introduction to the Physics of Granular Materials, 2000.

478832 The Effects of Additives on the Cold Flow Properties of Biodiesel

Abstract:

Biodiesel is a viable alternative to petro-diesel and has a number of advantages over it, but biodiesel has poor cold flow properties, most notably its high cloud point. The purpose of this research was to analyze the effects of dimethyl azelate on various fatty acid methyl ester (biodiesel) blends. Dimethyl azelate was chosen as it is formed as a by-product in a particular mechanism to form biodiesel and it has a similar structure to fatty acid methyl esters. The research shows that dimethyl azelate improved cold flow properties by lowering the cloud point. As the amount of dimethyl azelate was increased, the cloud point further lowered. The research also showed interesting results such as lower cloud points for binary fatty acid methyl ester solutions than the single components. These results are attributed to co-crystallization of fatty acid methyl esters of similar chain length. The co-crystallization was proved by cooling a biodiesel sample at a slow rate and noticing that similar chain length fatty acid methyl esters did not have separate peaks, but instead crystallized together.

Further work was dedicated to fitting the cloud point data to ideal solution theory, which assumes that the components of the biodiesel will not interact with each other. The research indicates that biodiesel with similar chain length fatty acid methyl esters does not follow the ideal solution theory due to co-crystallization of the similar chain lengths. Furthermore, as higher mass percents of dimethyl azelate were added to the biodiesel, the actual cloud point deviated further from the predicted cloud point. This suggests that some interaction between the fatty acid methyl esters and dimethyl azelate is occurring, but future research is needed to explain why.

478852 Biodiesel Production from Jatropha curcas: Experimental Investigation and & Process Simulation

Abstract:

Biodiesel production from Jatropha Curcas: Experimental investigation and & Process simulation

Bilal Ali Khan*, Javeria Ahmed, M. Shahab Alam
Dawood University of Engineering & Technology, MA Jinnah Road Karachi 74600, Pakistan
bilalalikhan465@yahoo.com, jiaahmed1@yahoo.com

Abstract

The challenges in continual depletion of fossil fuels and the relevant economic crisis and environmental concerns have attracted the scientists to seek for alternatives to fossil fuels like renewable energy resources. Biodiesel is an alternative fuel obtained from biomass and has similar physicochemical properties to petro-diesel. Jatropha Curcas (J.C) is one of the highly promoted oilseed crop at present in the world. The technological methods that can be used to produce biodiesel (fatty acid methyl ester) are esterification and transesterification with acidic and alkali catalysts. In this work, we took 55 ml of J.C oil (triglycerides) and 22 ml of methanol as solvent with sulfuric acid as catalyst (1 mole percent). The reaction was carried out at 55 °C for 1 hour. Similarly, 12 ml of methanol with sodium hydroxide catalyst were added and the sample was kept for 1.5 hour for reaction at 60 °C. In the reaction, 35 ml of fatty acid methyl ester along with 19 ml of Glycerol were obtained. For this purpose we also simulated the process for a prototype and compared results of simulation and experiment. The simulation results show a yield of 92%, whereas experimental data shows 68% yield. The difference in simulation and experimental results is attributed to the losses in experimental yield in laboratory.

Keywords: esterification, transesterification, fatty acid methyl ester (FAME), triglycerides

478910 A Promising Anhydrous Proton Transport Membrane: Hydroxyl Functionalized Graphane

Abstract:

Hydrogen fuel cells have high potential for becoming a major alternative to traditional combustion engines. An important component to these fuel cells include surfaces made of proton conducting material which allow the transfer of H^+ ions from anode to cathode at relatively high rates. Commercial proton transport membranes, such as *Nafion*, experience limited operation temperature due to the need for hydration. Graphane membranes functionalized with hydroxyl groups conduct protons at significant rates without a hydrated surface. Density functional theory calculations show that a 1-dimensional hydroxyl chain on graphane has a proton diffusion coefficient that is an order of magnitude higher than that of *Nafion*. We have investigated the transport properties of a membrane functionalized with a 2-dimensional network of hydroxyl chains that we proposed to be more prominent and capable of operating at higher temperatures than commercial membranes. We have performed molecular dynamics simulations on model surfaces using the *Vienna Ab-initio Simulation Package* (VASP). Our simulations occurred within a range of high temperatures (300K-1000K). Diffusion coefficients were obtained at each temperature, and the activation energy for proton transport on our membrane was determined.

479621 Constructing and Testing a Photobioreactor Designed for Precise Characterizations of Growth of Microalgae for Biofuel Applications

Abstract:

Algae biofuels are a renewable alternative to fossil fuels for energy production and associated carbon recycle. A common current practice for growing algae in a laboratory setting involves delivering a mixture of air and CO_2 through the top of a gyrating Erlenmeyer flask. This setup can lead to slow growth due to light penetration limitations. In this study, we designed, constructed, and tested a flat-panel channel-flow photobioreactor (named the Alginator) capable of facilitating rapid growth of high-density algae with a particular focus on light availability and defined circulation patterns to avoid CO_2 depletion or oxygen accumulation. The Alginator design provides a high surface area to volume ratio, allowing for consistent light penetration for the defined pathlength of ~1cm for the entire flat flow path. By sparging the gas mixture with small bubbles at the bottom of a riser for a loop-type air-lift configuration, the design increases CO_2 (and O_2) mass transfer while driving circulation of the bioreactor at a minimum gas flow rate.Â This minimization of flow and a defined exit at the top should permit off-gas analysis for measurements of CO_2 consumption and oxygen evolution.

Initial experiments were conducted to test the Alginator using Chlorella vulgaris in WFAMC media where growth of 6.7 mg/L was more than 10-times greater than achieved in a control shake flask culture (0.342 mg/L) due to extended exponential growth beyond an optical density of OD-550nm = 10. Towards evaluating energy and carbon capture efficiencies, Chlorella was grown on replete and â..." nitrogen-limited WFAMC media. Biomass was characterized by dry weight, hydrothermal liquefaction (HTL), bomb calorimetry, and optical density. As expected, N-limited growth was reduced considerably to 4.71mg/mL. This result clearly indicates that the Alginator provides for rapid growth of algae to high cell densities as a result of its design which will be the basis of comparing numerous other algae for

energy capture efficiency. Using HTL, the crude oil yield from the Alginator was 24 ± 1.5% in complete media and 18.3 ± 4% in nitrogen limited media. The lower yield in this preliminary test was unexpected, as we hypothesized that higher fatty acid content during nitrogen limitation would improve HTL yield. Further work is needed including our recent reconstruction of multiple Alginators where we will be able to better control temperature and more rapidly test different algae and growth conditions. A major goal of this work will be to provide a well-defined energy balance to assess energy and CO_2 capture efficiencies.

478197 Butanol Production from Corn Residue Using the Simultaneous Saccharification and Fermentation Method with Genetically Modified *Clostridium Cellulovorans*

Abstract:

Butanol production by fermentation of biomass, rather than by petrochemical routes, has regained interest due to butanolâ€™s strong potential as a biofuel which would not require modifications to current infrastructure. However, production by fermentation of glucose is not cost-effective. The use of lignocellulosic agricultural residue from corn, rather than nutritious corn product, would reduce substrate cost, as well as greenhouse gas emissions. *Clostridium cellulovorans,Â* genetically modified for overexpression of *adhE2,Â* was examined for use in the saccharification and fermentation of corn cob, corn fiber, corn bran, and corn husk. Pretreatment methods for the corn substrate were also examined, with each substrate being treated with sulfuric acid, calcium hydroxide, and sodium hydroxide. Bacterial growth was measured daily, and the composition of the supernatant, including butanol, ethanol, acetate, and butyrate, was tested using HPLC. It was determined that, for both bacterial growth and butanol yield, sodium hydroxide is the most effective pretreatment for lignocellulosic corn agricultural residue. For certain substrates and pretreatment methods, a butanol titer >2g/L was achieved in serum bottles in a ten-day fermentation.

479108 Palm Kernel Shell (PKS) of Oil Palm Waste Pyrolytic Bio-Oil for Bio-Fuel Products

Abstract:

The worldâ€™s demands on the limited natural resources that are used to power industrial society are diminishing as the demand rises. In complementing the petroleum industry to overcome worldwide energy crisis and to reduce greenhouse gas emission, bio-fuel has emerged as a compelling renewable energy source. In this paper, bio-oil is produced by pyrolysis method, a widely accepted feasible thermal conversion process in the absence of oxygen. The biomass used are palm kernel shells (PKS), one of the main oil palm tree residues which are now extensively sold as complementary fuel around the world. The process was performed in a microwave pyrolyzer at a temperature of 350 Â °C and a reaction time of 90 min with the use of 0.25-0.50 mm of ground PKS sample. The bio-oil produced is treated via distillation technique to yield both residue and distillate. This paper summarizes the characterization of PKS pyrolytic bio-oil before and after distillation (residue and distillate), and further determines similarities in terms of compositions between bio-oil and petroleum product which in this case is gasoline. Results obtained from FTIR, GC-MS and GC-FID analysis identified the existence of desired organic compounds, specifically benzene, xylene and phenol groups in PKS pyrolytic bio-oil. Acid value of the samples were also determined and the distillate shows low acidity which indicates low corrosivity of the samples. However, the calorific values obtained were much lower as compared to gasolineâ€™s due to high water content in the PKS bio-oil. It can be concluded that distillation of PKS bio-oil does improve the quality of the bio-oil but further modification or improvement on the distillation or treatment method needs to be done with the aim of reducing the water content of bio-oil.

Keywords: Renewable energy, Energy Crisis, Pyrolysis, Bio-oil, Palm Kernel Shell.

479714 Effect of Temperature on the Conversion of Triolein to Biodiesel Via a Heterogeneous Acid-Catalyzed Interesterification Reaction

Abstract:

Abstract

With the increase of environment protection consciousness and decrease of petroleum reserves, the search for alternatives energy has gained much attention. Biodiesel is a potential energy source that has grown in importance over recent years, as an alternative to petroleum based diesel fuel. Biodiesel has several benefits, such as reducing the greenhouse gas emissions. And it can be produced from varies of vegetable oils and animal fats, such as soybean and sunflower. The industrial production of biodiesel is by a transesterification reaction between triglycerides (found in oils and fats) and methanol in the presence of a catalyst. Stoichiometrically, 3 mol of FAME (fatty acid methyl ester) and 1 mol of glycerol are produced from every 1 mol of triglyceride. One of the main drawbacks of this process is the production of low-purity glycerol. Glycerol is limited in its application and has reached market saturation. Theoretically, glycerol by-production can be prevented by an alternative reaction, interesterification where triglycerides are reacted with methyl acetate instead of methanol. The stoichiometry remains identical but the replacement of methanol with methyl acetate produces triacetin as a byproduct instead of glycerol. Triacetin has been shown to have various positive applications after its purification. It can be used as a fuel additive and its mixture with FAME shows no detrimental effects on fuel properties. However, interesterification reaction has a lower reaction rate compared to transesterification, which limits the industrial applications of interesterification. In this project, triolein, a symmetrical triglyceride, was selected as the feedstock. Ferric sulfate worked as a heterogeneous acid catalyst to increase the reaction rate. The yield of methyl oleate/ethyl oleate via interesterification reaction was determined at different temperature with different combination of reactants to demonstrate the effect of temperature on the conversion of triolein to FAME. The highest yield of FAME was found to be approximately 70% at 120 °C with 20:1 MAOMR and 7.5 wt% catalyst loading.

479319 Ultrasonic Pretreatment of Microalgae for Enhanced Lipid Extraction

Abstract:

Oil extracted from algal biomass could provide a potential feed source for the production of bio-diesel to supplement current fossil fuels. Ultrasound pretreatment is a potential mechanical method to increase the yield of lipids extracted from the algae via solvents due to the cavitation breaking algae cell walls. A CCD was used to design experiments with varying power (350-750 W) and time (5-30 minutes). *Chlorella vulgaris/Cyanobacteria leptolyngbya* was harvested locally and pretreated at the powers and times specified by the CCD before extraction was done. Ultrasound pretreatment saw an increase yield up to 94% compared to the controls on non-treated algae suspension. Hexane (1:1 v/v) was evaluated as an alternative to the standard Bligh & Dryer (2:2:1.8 v/v/v chloroform/methanol/cell suspension). Yield was evaluated gravimetrically and Bligh & Dryer method yielded 134% greater yield than hexane. GCMS was used to evaluate composition of lipids after extraction. Lipids were found to be primarily composed of fatty acids and fatty acid methyl esters. SEM was used to analyze whether ultrasound had any effect on algae structure before and after treatment. Ultrasound proved to be a viable method in the pretreatment of algae for lipid extraction. Future research should be done into potential new solvents and finding the power and time that produced the maximum yield in lipids extracted.

479827 Structural and Functional Analysis of Spinels in Energy Storage Devices

Abstract:

As our world becomes increasingly mobile and wireless, our devices, vehicles, and homes demand more energy for longer amounts of time. With more research focusing on renewable energy generation, it is important to have high-powered and long lasting storage devices for optimum energy usage. The most recent electric storage technology introduced was the Lithium-ion battery in 1992 and since then, major efforts have been made to improve the most critical component: the cathode. Our research aims to study specific cathode materials called spinels at the nanoscale in order to fully understand what is necessary to optimize battery capabilities. Several spinels were synthesized with various elements in the crystal A site. The synthesis of the spinel cathode material was conducted via coprecipitation and then hydrothermal heating. The product was characterized by X-Ray Diffraction to confirm accurate crystallinity. The cathode was then made by creating a slurry of the spinel active material, carbon mixture, and a polymer binder. The slurry was coated onto Aluminum foil and dried in a vacuum oven. 1 cm diameter cathodes were punched from the coated foil and assembled into battery coin cells under inert gas in a glovebox, which were then discharged and recharged. Impedance testing was done to measure resistance within the batteries before and after each charge cycle. Our research also involved studying a new technique called Scanning Electrochemical Microscopy, which can be used for determining electrical conductivity of surface materials. SECM was found to be promising for measuring the conductivity of our various cathode materials while in electrolyte solution, which better simulates actual battery conditions. The results of this study found a promising cathode material for high-powered and long-lasting Lithium-ion battery applications. The methods used for synthesis proved efficient for making battery coin cells and the techniques used for characterization proved effective in forecasting battery performance.

479834 Impact of Flow Rate and Flow Field Design on Redox Flow Battery Performance

Abstract:

As the demand for electricity increases, a desire exists to improve grid efficiency and stability. Energy storage research can mitigate this need by providing steady power distribution to be used for peak shaving, power quality assurance, and more. Additionally, the intermittent forms of electrical production, solar and wind, can also be improved by providing a steady stream of power. One device that is a strong candidate for grid-level storage is a redox flow battery (RFB). RFBs have independent scaling of power rating and energy capacities, as well as having the active material dissolved in liquids active material. Most RFB literature focuses on identifying new active materials and electrolyte formulations, leaving cell engineering often overlooked. However, for RFBs to be viable, the energy conversion must be quick and efficient, enabled by low area specific resistance (ASR). Few systematic studies have been performed on these enhancements methods. This study investigates the performance variations by quantifying the effects of flow rates and flow field design on the mass transfer contribution to the total ASR. Single electrolyte flow cells, employing an iron (II/III) chloride electrolyte, enables this study. An equimolar solution of iron (II/III) chloride electrolyte is pumped to the cathode side of the flow battery and is reduced. The effluent is then transferred to the anode side and re-oxidized before returning to the electrolyte reservoir. We examine the interdigitated, flow through, parallel, and serpentine flow fields design with various electrolyte flow rates and active species concentrations. Potentiostatic and impedance measurements are taken to generate iR-corrected polarization curves, which are used to determine ASR. We find that ASR decreases as a function of flow rate for all flow field designs according to a nonlinear relationship. Our further studies have investigated modeling the experimental polarization curves. Furthermore, the performance of the interdigitated and flow through designs are more sensitive to changes in flow rates than serpentine or parallel designs. We performed an additional experiment in a 25 cm^2 single electrolyte cell to demonstrate how these performance trends scaling with increasing active area of the flow cell. With this study, an upper limit flow rate can be determined, as well as the best flow field design to enable the best overall flow cell performance.

480432 The Effects of Convection on Combustion of Nanoparticle Infused Fuels

Abstract:

In the fuel industry, there is an abundance of research being conducted to understand droplet combustion of nanoparticle infused fuels. Much of the previous research in this field focuses on how to understand general droplet combustion or how radiation affects droplet combustion, however little is being done to understand the effects of convection on the combustion of these infused fuels. The experimental set up currently used at the University of Iowa was modified to prevent radiation from entering the system and used in combination with a fan thus providing a method to continuously drive the combustion forward. In using this process, the droplet never ignites and as such burns only through convection. From this type of burning, the effects of convection were studied by determining the burning rate.

480433 Using Wasted Thermal Energy to Improve Home Appliances Performance and Reduce Electric Power Consumption

Abstract:

Using Wasted Thermal Energy to Improve Home Appliances Performance and Reduce Electric Power Consumption

Currently, Puerto Rico faces an energetic and economic crisis. The cost of kWh ranges from 26 cents to 29 cents; with the average household consumption of energy being about 908 kWh monthly. This means the average household in the island is currently paying about $263 a month on their electricity bill. While the Electric Power Authority of Puerto Rico sets the monthly cost per kilowatts; the amount consumed can be up to the client. Home appliances such as the refrigerator, heater and air conditioning unit are essential household equipment that tend to consume the most energy. Some household appliances, struggle to work within high ambient temperatures like those experienced in Puerto Rico. Refrigerators, for example, can't work properly if the ambient air around the condenser is not at least ten degrees lower than the condenser so that the energy can be given off as radiant heat. If the maximum temperature is exceeded the appliance will struggle to keep the temperature on the inside down and would run constantly or at least for longer periods of time. This could cause overheating, breakdowns and higher energy consumption.
After analyzing this problem, students of the Electrical Engineering Department of the Polytechnic University of Puerto Rico develop a Thermal Energy System that could help to reduce the electrical energy consumption. The objective of this research's work it's to improve the thermal system built by the electrical engineering students; in order to accomplish this, it was proposed to use the wasted thermal energy of the system to improve home appliances performance and reduce electric power consumption. The first phase of this research consisted on finding an optimum working fluid for the system and the second phase was to design a heat exchanger network that use the wasted thermal energy and install the same in the electrical engineering thermal prototype.
The electrical engineer students were originally using water as the system's fluid in order to cool down the inside prototype's temperature; as result they only drop down the same one °F at some hours of the day, but it wasn't constant all day and at some peak hours it was even higher than the control prototype. The chemical engineer students proposed a coolant as an optimum fluid. Two coolants were tested: ethylene glycol and propylene glycol at three different concentrations each. Experimentally was proven that ethylene glycol 50% reduced the temperatures inside the prototype two to three times more than only water and the effect was even through the whole day, so electrical appliances inside a house that has this system would consume less electrical energy. In the second phase, two heat exchangers were added so the heat removed from the prototype was transferred to a heater and air conditioning, in order to them use less electrical energy and work with wasted thermal energy as well. Scaling this work to run an average house could save people money on their monthly electric bill while, helping the environment with an ecofriendly house system.

480509 Catalytic Combustion of Bio-Diesel over Perovskite Catalyst

Abstract:

Heat energy sources are important not only for operating vehicles but also to warm up houses, cooking, and generate electricity from thermoelectric devices. Nowadays, the most world widely used heat energy source comes from burning fossil fuels which will run out in the near future. Therefore, finding a renewable energy source and combustion method, with high efficiency and reduced amount of harmful emissions, is necessary to protect our environment and thus lower the cost of energy. Biodiesel is a biofuel formed by a mixture of methyl esters with two or more oxygen atoms in their molecules, which makes easier to reach completed combustion, as compared with commercial diesel fuel. In addition, biodiesel can be synthesized from the biomass present in the waste of agricultural process, which reduces the costs of production as well as makes it an environmentally friendly process. Normal combustion of biodiesel occurs at very high temperatures (above 1500°C), and produces harmful pollutants such as nitrogen oxide, soot, and carbon monoxide due to uncompleted combustion. Expensive ruthenium-based materials have been used as catalysts for the combustion of biodiesel below 1,000°C. However, these noble-metal based materials are cost-prohibitive for commercial operations. In the present work, we have developed a lanthanum-manganese perovskite catalyst supported on yttrium-stabilized zirconia (LM-YSZ) as an inexpensive alternative to the aforementioned ruthenium-based catalysts and chemically stable in oxygen-rich environments. In the other hand, we tried lanthanum manganese oxide in different support such as titanium oxide. However, the experimental data obtained from our 24-h activity tests indicates that LM-YSZ shows a remarkable stability as well as a high carbon dioxide conversion than the lanthanum manganese catalyst in titanium oxide support where the oxygen carbon ratio is 1.6 compare with the stoichiometry ratio is 1.41. In addition, the mechanism of combustion in LM-YSZ is bi-function which means YSZ burn biodiesel and LM convert all carbon monoxide to carbon dioxide lead to completed combustion. Next, catalytic combustion in LM-YSZ is fuel flexibility which means it can burn completely not only biodiesel but also gasoline. Finally, Lanthanum Manganese oxide perovskite catalyst supported in YSZ showed high efficiently, cleanly and economically generate heat energy from catalytic combustion of biodiesel at the lower temperature.

479890 Modeling the Limiting Efficiency of Perovskite/Lead-Sulfide Quantum Dot Tandem Solar Cells

Abstract:

Multi-junction (â€œtandemâ€) solar cells have consistently achieved the top power-conversion efficiencies in the world for nearly three decades. In combination with solar concentrators and high-performance materials like GaAs, these devices routinely break the thermodynamic efficiency limit derived by Shockley and Queisser for single-junction devices. At the same time, quantum dots (QD) and Perovskites have become highly attractive as low-cost active materials, the former for their wide tunable band gap range and the latter for their high power-conversion efficiencies (PCE).

Here we model the limiting efficiency of a two-junction tandem solar cell with active layers of Perovskite material $CH_3NH_3PbI_3$ ($MAPbI_3$) and lead sulfide quantum dots (PbS QD). Starting from the ideal single-cell case (32.47% maximum PCE), we then consider a version of the PVMirror system proposed by Yu et al., with a long-pass dichroic mirror directing short and long-wavelength light onto the wide and narrow-band gap junctions, respectively. An idealized thermodynamic model was used to optimize the band gap of each junction and the splitting wavelength of the dichroic mirror. When the two junctions are wired independently, PCE is maximized with a splitting wavelength of 700 nm and band gaps of 1.77 eV and 0.94 eV for the Perovskite and QD junctions, respectively. This optimal system achieved a theoretical maximum PCE of 46.12%, with the Perovskite and QD junctions converting 27.80% and 18.32% of the AM1.5 spectrum, respectively. The predictive accuracy of the model was tested by comparing the theoretical and experimental performance of a tandem solar cell comprising 1.59 eV $MAPbI_3$, 1.08 eV PbS QD, and an 800 nm dichroic mirror in both independent and series wiring configurations. In the independent configuration, the tandem cell achieved a maximum experimental PCE of 18.08%, but this number was reduced to 6.45% when wired in series, indicating significant losses due to circuit resistances. By contrast, the ideal PCEs predicted by our model for

this particular system in the independent and series configurations were 42.99% and 33.96%, respectively. This discrepancy between predicted and measured PCE calls for further refinements to our model to account for losses due to realistic absorbance profiles and nonradiative recombination. Additionally, efforts should be made to improve electrical contacts between the two junctions in future experiments in order to ascertain the additional loss mechanism present in the series circuit.

479956 A Theoretical Study of Sulfur Adsorption on Ion-Exchanged Mesoporous Y Zeolites

Abstract:

A Theoretical Study of Sulfur Adsorption on Ion-Exchanged Mesoporous Y Zeolites

Zachary Blanchette, Kevin X. Lee, and Julia A. Valla

Fossil fuels contain sulfur, which releases SO_x when burned. The produced SO_x creates a major environmental problem as it contributes to acid rain and can harm the human respiratory system. To address this challenge, the Environmental Protection Agency (EPA) has released new regulations which require minimal amounts of sulfur in transportation fuels (diesel and gasoline). The conventional method to remove sulfur from transportation fuels is hydrodesulfurization, which requires high temperature and pressure and is incapable of removing refractory sulfur compounds. Adsorptive removal of sulfur using zeolite Y is a promising new technique for desulfurization. The introduction of mesoporosity in zeolites allow adsorption of refractory sulfur compounds at the internal active sites. Additionally, metals like Cu and Ce can be added to the zeolite extra-framework to increase selective adsorption.

Mesoporous Y zeolites were modified using desilication and surfactant assisted (SA) methods and were ion-exchanged with Cu or Ce to create the modified zeolites. Breakthrough experiments were conducted to determine the capacity for refractory sulfur compounds removal for the different metal-exchanged zeolites. The adsorptive capability of all zeolites tested was dibenzothiophene (DBT) > benzothiophene (BT) > thiophene (TP). CuSAY was shown the have the highest capacity for refractory sulfur removal due to strong π-complexation interaction.

A computational analysis for the metal-exchanged zeolites was also performed to investigate the interaction strength between the sorbate and sorbent. Adsorption of thiophenic compounds on metal-exchanged

zeolites was calculated using density functional theory (DFT) in Gaussian 09. For all calculations, an m062x method was used with 6-311G basis sets. The thiophenic compounds examined were TP, BT, and DBT. The metal-exchanged zeolite used was CuY. The adsorption capabilities of CuY were as follows: DBT > BT > TP. Additionally, the DFT results show that CuY exhibits a higher adsorption capability than the parent Y.

Literature

1. Lee, K.; Valla, J., Investigation of metal-exchanged mesoporous Y zeolites for the adsorptive desulfurization of liquid fuels. *Appli. Catal. B Environ.* **2017**, 201, 359-369

2. Wang, L.; Sun, Z.; Ding, Y.; Chen, Y.; Li, Q.; Xu, M.; Li, H.; Song, L. A theoretical study of thiophenic compounds adsorption on cation-exchanged Y zeolites. *Applied Surface Science.* **2011,** 257, 7539-7544.

3. Velu, S.; Ma, X.; Song, C. Selective adsorption for removing sulfur jet fuel over zeolite-based adsorbents. *Ind. Eng. Chem. Res.* **2003**, 42, 5293 – 5304.

4. Hernandez-Maldonado, A. J.; Yang, F. H.; Qi, G.S.; Yang, R. T. Desulfurization of transportation fuels by π-complexation sorbents: Cu(I)-, Ni(II)-, and Zn(II)-zeolites. *Appl. Catal. B Environ.* **2005**, 56, 111 – 126.

Acknowledgment

This work was supported by the American Chemical Society / Petroleum Research Foundation (PRF+ 55900-DN15).

480608 Catalyzed Deposition of CNTs on Carbon Felt for Flow Batteries and Fuel Cells

Abstract:

As the worldwide demand for energy increases, the need for clean, efficient energy as a substitute for the steadily diminishing and environmentally harmful supply of fossil fuels increases as well. However, the transient nature of the leading forms of clean energy, namely solar and wind, hinder their proliferation in the energy grid despite advances in harnessing their power. As such, reduction-oxidation (redox) flow batteries have gained significant attention for their ability to convert electrical energy into chemical energy for storage and later convert it back. In the charge state, the incoming electricity (generally from some energy source) powers the reduction of a species on one side of the cell and the oxidation of a different species on the other. To discharge, the current is reversed, the oxidized species is reduced back to its original concentrations and the reduced species is oxidized back to its initial levels, as the cell provides power to an appliance or power grid. One hindrance in a flow batteryâ€™s performance is low redox reaction kinetics due to poor carbon electrodes. This limitation is often overcome by stacking several layers of electrodes, which increases the material cost as well as the volume of the cell. Our lab group has developed a chemical vapor deposition process for growing multi-wall carbon nanotubes (MWCNTs) on the surface of a carbon electrode. This procedure increases the active area both for reaction and at the transport interface with the proton exchange membrane (PEM) while reducing the amount of electrode material necessary as well as the overall costs. However, carbon paper is fairly brittle and easily breakable. The goal of this project is to demonstrate that the more robust and durable carbon felt enhanced by cobalt catalyzed MWCNT growth can serve as a viable electrode in a redox flow battery. If successful, carbon felt could be explored as a substitute for carbon paper in other kinds of fuel cells, flow batteries, and other electrochemical devices.

480616 Towards the Development of a Polysulfide-Polyiodide Redox Flow Battery

Abstract:

As we move towards a reliance on renewable energy, low-cost energy storage systems will play a pertinent role in their integration into the existing power grid. Redox flow batteries (RFBs) are an electrochemical energy storage system that have recently received increased attention due to their separation of energy and power units, durability, and scalability. With current commercial RFBs having high chemical costs, efforts have gone into finding more cost-effective active species. In this work, we demonstrate preliminary steps in developing a polysulfide-polyiodide redox flow battery. Sodium polysulfide and sodium iodide are first characterized by cyclic voltammetry, showing a reduction potential of -0.40V and 0.48V vs Ag/AgCl respectively, giving a theoretical cell voltage of 0.88V. The species were shown to not react upon mixing and as such are appropriate for a full-cell without significant concern for separator design. The polysulfide-polyiodide system demonstrates cycling capability in constant current flow cell experiments, able to run for over 20 cycles. However, it was found that polysulfide demonstrates significant kinetic polarization. Thus, the need for a catalyst-modified electrode on the polysulfide side was made clear. Nickel polysulfide was shown to have the lowest overpotential out of the catalysts tested and is a promising catalyst for use in future full-cell testing. Additionally, graphene is tested as a catalyst for the polyiodine, showing 25% increase in capacity. Future work on the system will be performed to demonstrate a high concentration system using the catalysts demonstrated here.

480684 Analysis on the Feasibility of Oxidative Dehydrogenation of Ethane with Electrochemical Means

Abstract:

The combination of horizontal drilling with hydraulic fracturing has transformed the outlook of the U.S. chemical industry sector, as domestic shale offer plentiful low-cost alkane feedstock. Steam cracking of alkanes for the production of olefins is an energy intensive thermodynamically limited process, requiring approximately 180 MJ/ton ethylene produced and emits large quantities of criteria pollutants (such as NO_X and SO_X). The capital intensity of current steam cracking technology, along with non-attainment status in many current locations limits the expansion of the U.S. chemical industry sector. Process intensification approaches, which increase reaction efficiency and better integrate reaction latent heat are attractive, as such concepts could significantly improve chemical manufacturing energy efficiency. An attractive alternative being considered is oxidative dehydrogenation (ODH) of ethane, as the process is exothermic without thermodynamic limitation. Several ODH process concepts are under development, include fixed/fluidized bed and chemical looping concepts. However, the lack of modularity, cryogenic air separation, and oxygen carrier deactivation/costs associated with these ODH concepts still need to be overcome.

An attractive alternative being considered by Ohio University (OHIO), is an electrogenerative oxidative ODH (e-ODH) process, which utilizes solid oxide fuel cell (SOFC) technology with oligomerization and CO_2 capture, to convert ethane into a transportation fuel product. The e-ODH process potentially offers significant advantages over current ODH concepts, including co-production of chemicals and electrical power and process intensification through inherent oxygen separation/modularity provided by the SOFC platform. To assess e-ODH process potential, OHIO has developed an Aspen Plus® simulation for a 500 bbl/day system, to evaluate process efficiency and support a preliminary techno-economic study. For this simulation, pure ethane and air both enter the system separately. Ethane reacts with oxygen from air across an SOFC to form ethylene, which is then oligomerized utilizing a zeolite catalyst into transportation fuel product. This presentation will study results including estimates of process efficiency, product costs ($/GGE), and environmental emissions potential.

480587 PEDOT Encapsulated Feof Nanorod Cathodes for High Energy Lithium-Ion Batteries

Abstract:

Improved energy storage is the key to improving technologies such as electric vehicles, grid storage, and portable devices. In order to improve energy storage, improvements must be made to energy, power, safety, longevity, and sustainability. The most common battery type on the market today is intercalation, where ions insert between the crystal lattice layers. Conversion chemistries, where ions reversibly react with ions, have the potential to double or even triple the energy density of common intercalation chemistries used on the market today. However, poor cycling stability, low energy efficiency, and low power density have prevented conversion chemistries from being more widely used. The conversion cathode FeOF is a particularly high energy conversion cathode with very poor cycle life. By synthesizing PEDOT encapsulated FeOF nanorods, the PEDOT improves electronic connection and prevents side reactions. The capacity of this cathode material delivers a capacity of 560 mAh g^{-1} at 10 mA g^{-1} and an energy density of over 1100 Wh kg^{-1}. This is over twice the theoretical energy density of the common intercalation material $LiCoO_2$. The PEDOT coated FeOF maintains a capacity of about 430 mAh g^{-1} at 50 mA g^{-1} for over 150 cycles with a capacity decay rate of only 0.04% per cycle. This is still poor compared to intercalation chemistries, but it is half the decay rate ever reported among conversion cathodes.

480601 An Investigation on the Effect of Neutral Red Concentration on Voltage and Current Outputs for Microbial Fuel Cells

Abstract:

A major contributor to environmental pollution is the increasing quantity of sewage produced by growing human population centers. To address the need for alternative waste-management solutions, we report here the use of a yeast microbial fuel cell to break down complex carbohydrates found in sewage and convert them into electricity. The study utilized a microbial fuel cell set with two 250 mL glass half cells. The anode cell contained sucrose, neutral red, and the yeast species *Saccharomyces cerevisiae*. The cathode cell contained sodium potassium phosphate buffer. The yeast culture was made using 5 g of Fleischmannâ€™s yeast extract grown in one liter of deionized water supplemented with 20 g of sucrose. This yeast culture was continuously mixed for one week before usage. The working and counter electrodes were made of a carbon fiber cloth weave with a surface area of 50 cm^2 and placed in the cathode cell. The anode cell contained a silver/silver chloride reference electrode. The membrane used between the cells was Nafion 212, soaked in deionized water before runs. The anode cell was sparged with nitrogen to remove oxygen from the chamber and under these anaerobic conditions, yeast will readily undergo respiration to produce electrons which can be harvested in the microbial fuel cell. This is accomplished by the neutral red, which acts as an electron mediator transferring high energy electrons from the yeast to the anode electrode.

Three electrochemical cell characteristics were measured using a potentiostat: voltage, anode potential, and amperage. The neutral red concentrations tested were 1, 3, and 5 mM based on previously reported values in literature. From the collected data, the 1 mM neutral red run was able to generate the highest peak cell voltage of 0.25 V but all three runs leveled out at approximately 0.2 V after 72 hours. The anode potential for each concentration run was 9 V, with minimal difference between runs. The 1 mM run generated 50 pA, approximately 5 times as much as the 3 mM run. The cell voltage is consistent with previous literature, generally placing cell voltage between 0.2 and 0.4 V for neutral red yeast fuel cells, while the current data is lower than previously reported values. Based on our data, it appears that increased electron mediator concentrations has a negative effect on cellular respiration in yeast cells, leading to reduced overall cell voltage and current generation.

480697 The Effects of Algae Growth Stage on Biocrude Characterization from Hydrothermal Liquefaction Using Chlorella Kessleri Algae

Abstract:

Hydrothermal liquefaction (HTL) is a process that converts biomass into a biocrude oil in subcritical water. High temperatures and pressures are required to achieve the subcritical stage creating a highly reactive environment. This environment produces a higher biocrude yield, as compared to other conversion processes such as lipid extraction, because it breaks down carbohydrates, proteins, and lipids to produce the biocrude oil. A large advantage of the HTL process is that it eliminates the high energy and thus costly step of dewatering in pyrolysis because HTL is run in an aqueous environment. The products of high ash algae are an aqueous product, a solid product, and biocrude oil. *Chlorella kessleri* algae was grown in a controlled environment and harvested during the exponential, transition, and stationary growth phases of the algal growth curve. The different growth stages resulted in varying protein, carbohydrate, and lipid composition. The collected algae was loaded into an unmixed tubular reactor at 10 wt% concentration of algae in water. During HTL the reactor was maintained at 350 Â°C for one hour. The resulting biocrude was extracted and analyzed for its properties. The techniques used for this analysis include GC-MS, simulated distillation based on thermogravimetric analysis, and CHN-O analysis. This study analyzed the differences in biocrude oil properties for algae harvested during each of the different growth phases to see how this affected the resulting oil compositionâ€™s molecular, elemental, and fractional composition. The amount of unsaturated hydrocarbons increased by a total of approximately 6% moving along the growth phase and the amount of saturated hydrocarbons decreased by approximately 5%.Â The weight percentages of the lighter fractions of Heavy Naphta and Gasoline varied by approximately 5% among the different growth phases. The observed differences in biocrude composition suggest that to promote desirable oil characteristics algae for HTL processing should be harvested during the exponential or transition growth phases.

477717 Gas Oil Separation Plant Optimization in North Ghawar Producing Department

Abstract:

This report presents the work performed during COOP student assignment which was between: June 3rd â€" December 31th, 2015. The student received training with Saudi Aramco, Saudi Arabian national petroleum and natural gas company based in Dhahran, as Process Engineer for North Ghawar Producing Department in Abqiaq and this experience was very helpful especially for a fresh engineer like him. This report describes the job was done as well as some new topics were learned. it contains four case studies which were performed during the assignment namely: The Efficiency of High Pressure Production Traps (HPPT) based on the Actual Rate and Retention Time, Shedugm-5 HPPT Separation Performance Review, Disposal System Optimization and Hysys Simulation. Each case was discussed as well as its calculation and recommendation. In the first case study, it was found that the retention time of oil at high pressure production trap (HPPT) in Shedgum Gas and Oil Seperation Plant number 5 (GOSP-5) is 2.8 min which is below Saudi Aramco standard (above 4 min). Thus, the separation inefficient. In the second case study , we have discussed the inefficacity of the separation problem in Shegdum plant number 5. The reason of this problem were listed and discusses in detailed. Four solution were proposed and some of them are already exsist in Saudi Aramco facilities. the third case study was a simulation of the disposal system In Ain Dar area. Finally, the last case study is a simulation as well. However, it was conducted to check the amount of oil in water then compare it with Aramco standard.

478083 Development of High Performance Liquid Chromatography for Lanthanide Separation

Abstract:

Nuclear fission produces multiple lanthanide isotopes, many of which yield valuable insights into the type and origin of nuclear material. The ability to carefully separate and analyze selected lanthanide radioisotopes can be a key tool for nuclear forensic applications. However, adjacent lanthanides are notoriously difficult to separate from each other due to their very similar chemistry. One effective method for separation of lanthanide cations is High Performance Liquid Chromatography (HPLC). HPLC techniques have long been used to separate lanthanides, and our goal is to modernize our current capabilities through the development of a new test system. The chemistry is well developed, selective stripping of trivalent lanthanide cations from a cation exchange resin using Î±-hydroxyisobutyric acid (HIBA), but instrumentation advances yield scope for process improvements. Recently, a new HPLC system has been installed comprising of quaternary pumps, columns, fraction collectors and a Diode Array detector that allows for post column on-line analysis. Addition of PAR (4(-Pyridylazo) resorcinol)) for post column derivatization facilitates accurate determination of lanthanide cation elution using the diode array detector. The yellow to red color change upon lanthanide complexation is monitored through absorbance at both 414 and 530 nm. Initial work focused on determination, and then removal, of resin impurities that interfered with colorimetric detection. Single lanthanide cations were then run through the HPLC to â€˜mapâ€™ the system before moving to more complex mixtures. Finally, the impact of variations in HIBA concentrations and solution pH has been studied for a lanthanide separation of relevance to nuclear forensic applications, the separation of Eu, Sm, Tb and Y. Eu and Sm are adjacent lanthanide elements while Y behaves like a lanthanide element adjacent to Tb.

Â LA-UR-16-25242

478124 Functionalized Composite Silica Thin Film Membrane for Biomimetic Separations

Abstract:

Composite silica thin film membranes are proposed as a less costly and more efficient alternative for dilute aqueous solute separations. By immobilizing boronic acid carriers within the lipid bilayer in the silica thin film, selective transport of sugars can be achieved. However, the absence of boronic acid carriers does not prevent unwanted permeability of sugars and impurities through the membrane. Our current work focuses on investigating and minimizing this background leakage by tethering the lipid bilayer to silica thin film with the lipid-like silane 13-(trichlorosilylmethyl)heptacosane. This will form a more cohesive lipid bilayer and in general, a better membrane.

Research Supported by the National Science Foundation REU Program #EED-1460486

478211 Mg/DOBDC MOF Functionalized with Ethylene Diamine for Improved Resistance to Humidity in CO2 Capture Applications

Abstract:

With the steady rise in carbon dioxide levels, there has been an increased interest in various methods of separating CO2 from flue gas, as one of the major sources of carbon emissions is the burning of fossil fuels. Capturing CO2 before it reaches the atmosphere could mitigate its possible effects on climate change. Many materials currently used in CO2 separation consume a lot of energy and produce harmful by-products. Metal organic frameworks (MOFs) are crystalline structures combining metal ions with organic molecules, and they could provide a safer, less expensive alternative to separate CO2 from various gases. However, the effectiveness of MOFs for CO2 capture decreases along with the degradation of their structure in the presence of humidity, which is another major component of flue gas.

To minimize the effects of humidity to the structure and CO2 capture performance of MOFs, this work investigates the addition of amine functional groups to improve the stability of the MOF structure, using Mg-MOF-74. Amines are currently used as a chemical absorbent in CO2 capture. Mg-MOF-74â€™s properties were first examined using x-ray diffraction and BET surface area analysis, as well as CO2 capture tests before and after steam treatments. This MOF was then synthesized and functionalized with amine groups at open coordinate sites in the hope of improving CO2 capture and protect the MOF from degradation. It was found that the modified MOF had improved CO2 capture, and was more resistant to degradation in the presence of humidity.Â Strikingly, the MOF showed no loss in capacity after extreme steam treatments when functionalized with amine groups, suggesting that the addition of amines could successfully mitigate the effects of humidity on the MOF structure.

478564 Adsorption Properties of Doxycycline Antibiotic from Aqueous Solutions Onto Pristine, Activated and Magnetized Carbon Nanotubes

Abstract:

Abstract

The presence of antibiotics in aquatic environments has been recognized as an issue warranting consideration. Increasing concerns have been raised regarding the potential risks of antibiotics to human and ecological health due to their extensive use. More efficient and cost-effective adsorbents are urgently needed and are the focus of intense research effort. Previous studies have demonstrated that carbon nanotubes can be used as effective adsorbent materials to remove organic contaminants [2].

In this study, magnetic multi-walled carbon nanotubes (MMWNTs) were synthesized from iron salt solution [3]. Pristine, oxidized and magnetized MWNTs were characterized by scanning electron microscopy (SEM), Fourier transform infrared spectroscopy (FT-IR), and X-ray diffraction (XRD). The three solid samples were then investigated for the adsorption of doxycycline antibiotic (DOX) from aqueous solutions. The influence of adsorbent mass, contact time, the temperature and the pH on the adsorption of DOX on the three types of solids was investigated by conducting a series of batch adsorption experiments. The equilibrium data at different temperatures were fitted with the Langmuir, Freundlich, Tempkin, Redlich-Peterson models. Adsorption results showed that the maximum percentage removal of DOX from aqueous solution by pristine, oxidized and magnetized MWCNTâ€™s were 99.39%, 78.79% and 94.68% respectively.

Our results indicate that surface properties and aqueous solution chemistry play important roles in DOX adsorption on MWCNTs.

Key words: adsorption, pristine multi-walled carbon nanotubes, oxidized multi-walled carbon nanotubes, magnetized multi-walled carbon nanotubes, water treatment, antibiotics, doxycycline, multi-walled carbon nanotubes, nanotechnology.

References

[1] Mauter, M.S.; Elimelech, M. Environmental applications of carbon-based nanomaterials. Environ. Sci. Technol. 2008, 42, 5843â€“5859.

[2] Liangliang Ji et al., Adsorption of tetracycline on single-walled and multi-walled carbon nanotubes as affected by aqueous solution chemistry. Environmental Toxicology and Chemistry. 2010, 19 : 2713-2719

[3] Fazelirad H, Ranjbar M, Taher M, Sargazi Gh Preparation of magnetic multi-walled carbon nanotubes for an efficient adsorption and spectrophotometric determination of amoxicillin. Journal of Industrial and Engineering Chemistry. 2015, 21: 889â€“892.

Acknowledgements

This work was supported by the Research Grant of Nazarbayev University.

477829 Graphene Oxide Quantum Dot Enhanced Nanofiltration Membranes for Water Purification

Abstract:

Abstract:

Graphene Oxide Quantum Dot Enhanced Nanofiltration Membranes for Water Purification

Nanofiltration membranes are used in many separation processes including food processing, drug synthesis, and water desalination. These membranes selectively exclude particles through size exclusion through the pores and electrostatic repulsion via the charged surface of the membrane. This research seeks to enhance the transport and antifouling properties of these membranes by modifying them with graphene. Graphene is a single hexagonal lattice of sp2 hybridized carbon atoms which has many desirable mechanical, optical, and electrical properties. Graphene quantum dots (GQDs), are small (less than 30nm) sectors of graphene which have carboxylic groups all around their periphery. GQDs can be attached to the surface of a nanofiltration membrane through EDC/NHS coupling, The aim of this research is to utilize the abundance of these electron dense carboxylic groups to add a negative charge to the surface of a Nanofiltration membrane. Our experiments developed a method for aminating the surface of a nanofiltration membrane and then attaching GQDs to this surface. Decrease in zeta potential and increased contact angle (gain of approx 15°) indicated that GQDs were attached to the surface of the membrane. This method was unsuccessful, however, in completely coating the surface, and future work will go toward a method that results in higher loading of GQDs to the surface of the membrane.

The next study seeks to create a selective membrane using the layers between graphene particles. This is done by depositing a layer of Graphene Oxide (GO) or GQDs onto a microfiltration membrane (PS35) then reducing the GO to inhibit dispersion into water. This membrane will be used as a support to a nanofiltration membrane under the active layer. The water flux, salt rejection, and mechanical strength of this type of membrane was tested and compared to Nanofiltration membranes.

478728 Development of Structured Packing Via Additive Manufacturing

Abstract:

Our research explores the role that additive manufacturing can play in the design and production of structured packing in an effort to improve the efficiency of small scale chemical laboratory equipment (e.g., condensers, distillation heads). Structured packing forces fluids to take intricate paths across their medium creating turbulent flow which leads to enhancements in mass and heat transfer rates. Furthermore, developments in additive manufacturing have led to unprecedented levels of specificity being achieved in the production of small scale devices and offer a promising medium both in terms of materials used and precision.

This study focuses on the design and testing of metal structured packing created through a CAD medium with an emphasis on creating surfaces with high surface area to volume ratios. After the production phase, these designs will be tested in a distillation experimental setup to measure the efficiency gains that these novel devices can offer over traditionally manufactured structured packing in an effort to further understand the potential application that additive manufacturing can play in small scale distillation experiments. We are particularly interested in developing â€œdrop-inâ€ packings that can be used with existing, commercially available laboratory glassware.

478819 Nanostructured Membranes for Water Application

Abstract:

The growing need for freshwater has lead to the water crisis we face today. Research into water purification to increase the availability of freshwater for various industrial needs and potable use has become focused on polymeric membranes. Polymeric membranes are flexible and have desirable properties for the purification of water and desalination. Incentives for researching interactive membrane filters are lower cost and lower energy consumption compared to traditional membranes. Polymer membranes also reduce the amount of contamination due to lower rates of degradation and chemically inert materials. Polyvinylidene fluoride (PVDF) membranes functionalized with polyacrylic acid (PAA) hydrogels, networks of crosslinked polymer chains, were used to remove heavy metal ions from a water source. Pretreated with Na^{+} ions PVDF-PAA membranes were loaded with Pb^{2+} and Ca^{2+} in ion exchange experiments.

478831 Tailoring the Hydrophilicity of Electrospun Membranes for Water Filtration

Abstract:

Saline water provides an alternative source to freshwater, which is becoming scarcer as supply dwindles and demand increases due to industrialization and an increased global population. Desalination membranes, particularly those in reverse osmosis, are susceptible to fouling and scaling, which increases monetary cost through energy, maintenance, and membrane replacement. Pretreatment membranes are used to reduce the concentration of foulants and scaling precursors in saline water, improving membrane lifetime and water flow through the RO membranes while reducing operation costs. Key design parameters for these pretreatment membranes are high fluxes and removal of fouling/scaling components. Electrospinning polymer membranes is an easily scalable process that produces non-woven, micron-to-nanometer scale fibrous membranes with tunable pore sizes (related to flux) and a high surface area for interacting with feed streams. A large body of literature precedents has demonstrated that increasing the hydrophilicity of polymer membranes reduces the propensity of fouling. Therefore, this project examines the combination of hydrophobic poly(vinyl chloride) (PVC) with hydrophilic polymer poly(vinyl alcohol) (PVA) to form porous, electrospun membranes with tailored hydrophilicity. PVA was chosen because it is highly hydrophilic, but displays a limited solubility in water under normal process conditions (e.g., <60 ĚšC). Polymer solutions of PVC/THF and PVA/water were characterized by solution rheology, and electrospun fibrous mats were prepared using side-by-side and layered electrospinning techniques. The fibers were analyzed using scanning electron microscopy (SEM) in order to determine their pore size and fiber diameter. Finally, the relative hydrophilicity was tested via water contact angle measurements. Our results indicate that the fiber diameters of both the PVA and PVC fibers are correlated to the solution concentration and viscosity, which has a subsequent impact on the membrane porosity. Additional experimental factors such as voltage, flow rate, and tip to collector distance have been used to tailor the fiber diameter and porosity.

478688 Recovery of Dilute Aqueous Butanol By Membrane Vapor Extraction with Dodecane or Mesitylene

Abstract:

A novel isothermal, nonselective-membrane separation process, designated membrane vapor extraction (MVE), efficiently recovers butanol from a dilute aqueous solution, for example, from a fermentation broth. In MVE, the two feed and solvent liquids are not in contact; they are separated by vapor. Therefore, compared to conventional extraction, MVE avoids formation of difficult-to-separate emulsions.

In MVE, a semi-volatile aqueous solute (e.g., butanol) vaporizes at the upstream side of a membrane, diffuses as a vapor through the membrane pores, and subsequently condenses and dissolves into a high-boiling nonpolar solvent, highly favorable to the solute but not to water. Design analysis of an industrial-scale MVE unit for processing 2-wt % aqueous butanol by dodecane solvent at 40 °C indicates over 90 % recovery of the feed butanol with essentially no water loss and with very low energy requirement. However, the published design study gives no experimental evidence for the calculated MVE separation.

Here, we present experimental data to validate the MVE process. We use an omniphobic (i.e., hydrophobic and oleophobic), 0.2-μm pore-diameter Versapore R membrane housed in a small 6-cm wide by 10-cm long plate-and-frame channeled flow cell with 0.8-cm gap thickness. The membrane flow cell is operated between recirculating flow loops.

Since vapor transport across the membrane contributes minimal resistance, MVE performance is governed by mass transfer through feed and solvent boundary layers. Prediction of the experimental-extraction performance determined from Graetz-Lévêque analysis agrees well with experiment using no adjustable parameters. Preliminary analysis of downstream solute recovery from the extract via distillation is shown to be more efficient than that for pervaporation

because of insignificant water carry over through the MVE membrane.

479045 Crystallization and Phase Transformation Kinetics of Suberic Acid Particles Formed Via Monodispersed Droplet Evaporation

Abstract:

In order to effectively deliver an Active Pharmaceutical Ingredient (API), pharmaceutical companies rely on specific particle properties including size, morphology, and structure to achieve the desirable drug performance. The ability of a solid to exist in more than one arranged crystal structure is called polymorphism. When polymorphism exists for a given compound, there is typically one stable polymorph and one or more less stable or metastable polymorphs at a given temperature and pressure. Alternatively, amorphous solids, solids without an ordered structure, are inherently less stable and can be induced with the help of an excipient. Less stable amorphous structures and metastable polymorphic structures typically crystallize first, followed by rapid recrystallization into the stable form. Methods can be used to preserve less stable structures by preventing the rapid transformation. However, the length of time these less stable structures are maintained for before they transition to the more stable structure is less defined. Internal structure is extremely important in pharmaceutical development for a variety of reasons. Less stable amorphous structures and metastable polymorphic structures are more soluble in the body and therefore have a higher level of bioavailability, the fraction of drug that dissolves appropriately in the patient and is delivered to the system to provide relief of symptoms. Also, it is important to know the entire polymorphic space before manufacturing pharmaceuticals to avoid future issues with either production or patents. In addition, the phase transformation from one polymorph to another can dictate drug shelf life. Thus, methods to identify and understand polymorphs are critical for success. In this work, we utilize a process of monodisperse droplet evaporation technology to form and further investigate less stable structures that are not obtained with identical chemistry using other formation processes.

Monodisperse droplet evaporation technology is implemented via the Vibrating Orifice Aerosol Generator (VOAG), which creates droplets of solution that are well dispersed from one another allowing for the solvent to rapidly evaporate. The rapid evaporation in the VOAG leads to high supersaturations, which drives a potential for formation of less stable structures. In this research, suberic acid is utilized as the solute or model pharmaceutical ingredient. In order to investigate the morphology, Scanning Electron Microscopy (SEM) is used. X-Ray Diffraction (XRD) and Differential Scanning Calorimetry (DSC) are used to analyze structure and understand phase transformation kinetics. In this work, we will present the formation of amorphous and metastable structures of suberic acid at conditions far from previous work. In addition, we will present the phase transformation kinetics of amorphous and metastable suberic acid. Thus, this work demonstrates the use of the VOAG as an effective method to investigate polymorphism.

479526 Separation of Racemic Penicillamine on Chiral Gold Nanoparticles

Abstract:

Penicillamine is a chiral active pharmaceutical ingredient. The D-enantiomer of penicillamine, sold under the trade name Cuprimine, is used to treat Wilsonâ€™s disease and rheumatoid arthritis. The L-enantiomer of penicillamine, however, is toxic. The enantiomers of racemic penicillamine must therefore be separated and purified before D-penicillamine can be put into a dosage form. The current method for enantiomeric separation used in the pharmaceutical industry is chiral column chromatography, which can be expensive and cumbersome. To avoid this cumbersome method, this work presents an innovative method for enantiomeric separation of racemic penicillamine using chiral gold nanoparticles. Spherical gold nanoparticles of diameter 4 nm are chemically synthesized. The gold nanoparticles are then coated in either L-cysteine or D-cysteine to render them chiral. The enantioselective interactions of these nanoparticles are then inspected using optical rotation measurements. Results demonstrate that when racemic penicillamine is added to gold nanoparticles coated in either D or L-cysteine, one enantiomer of penicillamine is selectively adsorbed onto the gold nanoparticles, leaving an excess of the other enantiomer in solution. Gold nanoparticles coated in D-cysteine adsorb D-penicillamine, leaving an excess of L-penicillamine in the solution phase. D-cysteine on gold nanoparticles has an initially positive optical rotation. As the amount of L-penicillamine (negative optical rotation) in the solution phase increases with subsequent additions of racemic penicillamine, the optical rotation decreases. Conversely, gold nanoparticles coated in L-cysteine adsorb L-penicillamine, leaving an excess of D-penicillamine in the solution phase. L-cysteine on gold nanoparticles has an initially negative optical rotation. As the amount of D-penicillamine (positive optical rotation) increases with subsequent additions of racemic penicillamine, the optical rotation increases. Control experiments indicate that D-cysteine and L-cysteine in solution show no net separation of racemic penicillamine, demonstrating that the gold nanoparticles are necessary for the enantiospecific selection to occur.

479771 A 3-D Printed Scaffold for Improving Coating Efficiency in Electrospun Membranes for Membrane Distillation Applications

Abstract:

This work focuses on the optimization of an existing initiated chemical vapor deposition (iCVD) process used for coating electrospun nanofiber membranes. Many of the more easily spinnable fibers are inherently hydrophilic, making them unsuitable for membrane distillation experiments. In MD, only water vapor is allowed to pass through a semi-permeable membrane while liquid water and salts are rejected. Chemical vapor deposition can be used to render these membranes hydrophobic and suitable for MD. Previous methods for coating these membranes involved securing the membrane to be coated parallel to the direction of the reactant gas flow, relying on the diffusion of reactants into the membrane to achieve coating. After coating of one side, the membrane would be flipped to repeat the procedure for the other side, demanding lengthy deposition times to ensure complete coverage. To improve this process, both in terms of coating time and reactant usage, a 3-D printed scaffold was designed that supports the membrane such that it is now normal to the gas flow, significantly increasing the reactant flux passing through the membrane itself, and taking advantage of the membraneâ€™s permeability. This configuration allows polymerization on the forward facing surface of membrane, while also promoting penetration through the pores to polymerize on the rear face as well. After several design iterations, the scaffold was successfully able to cut the deposition time by 75% without sacrificing the integrity of the coating. The time required to achieve sufficient coatingâ€"ignoring the time lost shutting down the process to flip the couponâ€"has been reduced from 100 minutes per side to 50 minutes in total without ever stopping the coating process. Successful membrane distillation experiments have shown that systemic coating occurred throughout the nanofibers. Uncoated PAN membranes will wet out during MD experiments resulting in salt crossover. The coated membranes exhibited 100% salt rejection indicating complete coating of the hydrophobic polymer during the CVD experiment.

479879 Removal of ^{210}po from Groundwater for Nuclear Astrophysics Applications

Abstract:

At the frontiers of physics research, searches for solar neutrinos, dark matter, and double beta decay have become more prevalent and the need for low background experiments has increased. New technology and equipment necessitate higher levels of sensitivity and the ability to extract weaker signals amidst high levels of background. The Borexino detector at the Laboratori Nazionalli del Gran Sasso (LNGS) in Italy is one such experiment which requires extremely pure liquid scintillator with low levels of radioactive isotopes in order to study solar neutrinos. The detector uses a liquid-liquid extraction system to purify the scintillator by removing polar impurities, but the naturally occurring radioactive isotope ^{210}Po is abundant in the groundwater used and not fully removed by the extraction. This motivated the development of a fractional distillation apparatus at Princeton University to remove compounds of ^{210}Po from groundwater, which would then be put in place at LNGS to use with Borexino. The apparatus consists of two distillation columns working in conjunction to remove the more volatile polonium hydrides followed by the heavier and less volatile dimethyl polonides; this results in purified water which would have lower concentrations of ^{210}Po. In order to determine the amount of polonium in the purified water, a precipitation reaction was run to create measurable samples and deposit the polonium onto thin disks of silver foil. A silicon surface barrier detector was used to count the alpha-emittance of samples, first for the well water samples and polished silver foils, followed by the purified water samples. Upon analyzing the data generated by the alpha radiation hitting the detector, it was possible to determine the presence of polonium in the groundwater and the effectiveness of the distillation column. While all samples had high levels of low-energy background, it was possible to see a reduction in the high-energy ^{210}Po decay peak after the purification process. The successful reduction of noise from ^{210}Po would allow for more sensitive detection of signals in such a low-background experiment. Further studies will determine the optimal reflux ratio resulting in the most efficient purification process while scaling up this distillation apparatus to the volumes necessary for LNGS and Borexino.

480446 Improving the Performance of a Liquid-Liquid Extraction Unit By Using Optimum Solvent

Abstract:

Improving the Performance of a Liquid-Liquid Extraction Unit by using Optimum Solvent

The Polytechnic University of Puerto Rico stopped the use of the Liquid-Liquid Extraction (LLE) Unit at the chemical engineering laboratory because it threatened the health of the operators and people near the area. The reason of this issue was the solvent used with this particular equipment, because it is hazardous to health, emits a strong odor and also causes corrosion in the equipment. Therefore, substituting this toxic solvent became a priority.
The LLE is a mass transfer operation unit based on the dissolution of one or more components of a liquid mixture in a selective solvent. This process consists in putting together a liquid mixture with a liquid solvent which is immiscible with this mixture and, by means of a good agitation, the solvent is put in contact with all the components of the mixture and selectively extracts the components that are more soluble in it than in the mixture. LLE has many applications in different industries such as: metallurgical extractions, pharmaceutics, fuel oil extractions, and environmental industry. It is an important process of the chemical engineering industry and if the students can learn more about it, the better; that is why this project is so important for the community and the institution. In this research, an optimum non-hazardous solvent to be used in the LLE Unit was found. Moreover, three Standard Operating Procedures (SOP) of LLE experiments were designed using this safe solvent.
The first phase of the research were to analyze the common solvents available to be used with this liquid-liquid extraction unit, and it was concluded that none of these solvents were suitable for solving the actual problem. All of them were dangerous for the health, and also had a low permitted exposure limit, established by regulatory agencies, for an operator to handle them without a fume hood. Because of this matter, the next phase of the project was searching for non-common alternative solvents to perform the experiment without harm. Green solvents are those non-toxic, biodegradable, reusable, and easy to separate solvents that can be used in industrial procedures without affecting human health or environmental safety. After a rigorous research, an optimum solvent was found in this category, meeting all the required specifications.
In the final phase, three new experiments were designed with their respective SOPs, in order to test the selected solvent. It was proposed to work at the following concentrations: 50% chosen solvent, 40% water and 10% solute. The selected concentrations were the appropiate ones, based on the literature review, in order to prevent any damage to the equipment and promote a good extraction. As future work, the validation of each SOP is required to confirm that the proposed concentrations are the most favorable.

479992 Thermodynamic Analysis of 1-Cyclhexylpiperidine for SPS FO Water Treatment Processes

Abstract:

Tertiary Amines have been found to switch polarity upon reaction with CO to form highly concentrated solutions that can be utilized to draw fresh water from high osmotic pressure mixtures such as seawater. By studying the 'switch' (gas contacting and degassing reactions) of these switchable polarity solvents we can aid in the optimization and scale-up of SPS FO water treatment processes. 1-cyclohexylpiperidine (CHP) has been found to produce up to 70 wt% solutions with an osmotic pressure over 500 atm. Gas contacting experiments on this species were performed at variable temperatures and pressures of CO_2, and concentrations were measured through freezing point osmometry. Results show that 1-cyclohexylpiperidine produces higher concentration solutions at lower temperatures and higher partial pressures of CO_2. The rate of draw solution formation decreases as temperatures and partial pressures of CO_2 go down. It was determined that optimal temperatures for these reactions were greater than around 77 °C for degassing and less than 51 °C for gas contacting operations. By deriving an equilibrium equation that models this data and performing a van't Hoff analysis, the values of $\Delta H°$ and $\Delta S°$ were determined to be -68.9 kJ/mol and -212 J/mol/K respectively. Some general kinetic implications can be garnered from this data, but further experiments are suggested to verify these kinetic models and obtain accurate rate constants.

480308 Seawater Desalination and Industrial Wastewater Recovery Using Hybrid Forward and Reverse Osmosis at the Pilot System Level

Abstract:

Forward osmosis (FO) is an emerging membrane technology that offers certain advantages over currently implemented forms of water treatment, reuse, and desalination. When coupled with reverse osmosis (RO), an economically viable technology solution can unfold as a hybrid FO-RO system. In this, FO acts as a pretreatment to RO while RO regenerates the spent draw solution and produces fresh water. For several years now, bench-scale tests have been done on this type of configuration, but information on module scale performance, system scale-up, system dynamics, effects of operating conditions, and recovery rate is lacking. In order to evaluate these parameters, a pilot-scale system with commercial FO and RO membrane modules is assessed. Performance was assessed for each system (RO and FO) individually and while run in tandem using similar operating conditions. From initial tests, we find no significant difference between the two setups in terms of osmotic performance, rejection, and recovery. From FORO operation, it has been found that feed volumes for both FO and RO systems need to be replenished quickly at high recovery rates. It was also seen that operating the RO system at low pressure self regulates overall dynamics of the system. Further, it was observed that high permeation of solute in either the feed (in FO) or permeate (in RO) limits long time closed loop operation. Future work will focus on the self-regulating nature of FORO, the effect of recovery on overall system performance in terms of rejection in RO and solute flux in FO, the effect of draw solute type, concentration, flowrate etc., the effect of module type, and a side by side comparison with pilot scale to bench scale systems.

480553 Defluoridation of Drinking Water through Adsorption Using Naturally Occurring Zeolites

Abstract:

Excessive fluoride intake can lead to a disease known as fluorosis which is characterized by decaying and weakening of teeth and bones. In certain areas of the world, such as Ethiopia, water sources are naturally at high fluoride levels. The most common methods used to remove fluoride from drinking water are reverse osmosis and activated alumina, both of which require costly resources and electrical power. For a developing nation such as Ethiopia, these processes are impractical. Our research seeks to find a way to use naturally occurring zeolites found in Ethiopia to defluoridate the water. The goal of our project is to determine the operating conditions, zeolite types, and zeolite amounts for fluoride removal in a flow process for applications in defluoridation from drinking water in developing regions of Ethiopia.

In our research, we selected two naturally occurring zeolites to study; analcime and mordenite. In prior research, these two zeolites demonstrated a capability to adsorb fluoride effectively. The mordenite was received calcined in powdered form while the analcime required grinding, sieving, and calcination. We characterized the zeolites using X-ray Diffraction (XRD) and Scanning Electron Microscopy (SEM) in order to understand their structures and morphologies and compare them to what is found in the literature.

Our fluoridated water came from the dilution of 0.1M fluoridated water to concentrations of 100 mg F^- per liter and 20 mg F^- per liter. The fluoride concentrations were measured using an ion-selective electrode. Our fluoridated water received additional treatment with pH adjustment using citric acid. The pH was adjusted to lower the acidity to a range of values for certain experiments. The pH was measured using a pH electrode.

Thus far, we have conducted several batch experiments to test the defluoridation capacities of the zeolites under varying fluoride conditions and pH levels. The pH was found to have a significant effect on the defluoridation capacity, with fluoride removal exceeding 90% under the following conditions:Â pH of 2.44, initial F^- concentration of 100 mg/L, and adsorbent dose of 100 g/L. Both analcime and mordenite show promise for providing an economically viable option for the removal of fluoride from drinking water in Ethiopia.

Acknowledgment

This research has been funded in part by the UConn Office of Undergraduate Researchâ€™s IDEA Grant program.

480588 Applying Basic Chemical Reaction to in-House Water Treatment

Abstract:

In 1974, the Safe Drinking Water Act was passed by Congress. This act detailed maximum concentrations of analytes in water such that the water is safe for consumption. Within this act, the maximum lead concentration before the EPA is required to intervene is 15 ppb. This concentration was surpassed in Flint, Michigan, after city officials changed water supply in mid-2014. One year later, a tap water sample sent into the EPA revealed a lead concentration of 104 ppb. It is believed that the lead leaching was a result of corrosive water passing through the pipes. Until the pipes can be replaced, the residents of Flint must either use bottled water or purchase an NSF-Certified water filter, which cost anywhere from $60-$500 new with replacement filters which must also be NSF-Certified. This project aims to give an alternative filtration system which utilizes precipitation chemistry to extract lead from water and engineering to give the most cost-effective separation method.

480744 Fabrication of Non-Fluorinated Hydrophilic-Oleophobic Stainless Steel Meshes for Oil-Water Separation

Abstract:

Oil-water separation is a worldwide concern due to the increasing emissions of oil-contaminated industrial water, frequent oil spill accidents and the shortage of drinking water. In this study, hydrophilic/oleophobic stainless steel (SS) meshes which can separate oil from water were fabricated by one step solution based coating method using methyltrimethoxysilane (MTMS). The coating materials were prepared by mixing MTMS with hydrochloric acid and sonicated in an ice bath for 5 min. Meshes were then immersed in the mixture for 2 min and dried in air. The dimensions of meshes were varied to study their effect on wettability and separation efficiency. Coated meshes were then used for gravity-driven oil/water separation with oil-water mixtures and oil-water emulsions. Contact angles, fluid flux, and separation efficiency were evaluated to determine the optimum mesh dimensions. The so called mesh 400 with an inter-fiber gap of 38 Î¼m and fiber radius of 13 Î¼m showed the best performance in separating oil-water mixtures. The coated mesh allows water flux as high as 7.16×10^4 $L \cdot m^{-2} \cdot h^{-1}$, with water recovery rate up to 91.1 % and separation efficiency higher than 99.9 % solely driven by gravity. XPS and ATR-FTIR were performed on coated stainless steel surface to prove the presence of silanol groups that play a critical role in the unique wetting properties. Even with the presence of sand particles on the surface after sand impact testing, the treated meshes were still able to separate oil-water mixtures with high separation efficiency and water recovery. The coated mesh filters hold great promise for applications in fields such as oil spill cleanup, where sand and dust will inevitably contaminate the system. Additionally, the ability to separate oil droplets larger than its pore size from oil-water emulsions makes the meshes a cheap, attractive candidate for pretreatment in current large scale oil-water separation units.

480784 Textile Wastewater Treatment By Bioinspired Antimicrobial Nanofiltration Membranes

Abstract:

Surface functionalization with advanced nanomaterials offers tailored control and targeted design of surface properties optimal for separations. Such modifications endow materials with improved or new separation qualities such as high hydrophilicity, excellent selectivity and permeability, and enriched antimicrobial activity. In this study [1], we develop two strategies (two-step deposition and co-deposition) that use mussel-inspired polydopamine (PDA) to strongly immobilize copper nanoparticles (CuNPs) onto a porous polymeric membrane. Through bridged surface cavities from ultrafiltration (UF) to loose nanofiltration (NF), these membranes display promising potential for the separation, recycle, and reuse of dye/salt mixtures as resources rather than as environmentally harmful waste. To confirm the optimization of membrane surface properties, a series of materials characterizations was executed: SEM, EDS analysis, AFM, water contact angle, and zeta potential measurements. The results indicate an overall high performance of surface properties with a homogeneous nanoparticle distribution, low roughness, favorable hydrophilicity, and relatively neutral charge. Co-deposition of PDA andCuNPs was verified as a facile and time-saving process that expedited a higher CuNP loading compared to the traditional two-step strategy, as confirmed by SEM and AFM images. The integration of polyethylenimine (PEI)-modified CuNPs with high density of positive charges was shown to play an important role in fine-tuning the hydrophilicity and compatibility with PDA and in largely neutralizing the negative charge of PDA, thus promoting an outstanding salt permeation (82% Na_2SO_4, 98% NaCl). In addition, CuNP/PDA-modified membranes performed an ultra-high rejection (>99.0%) of three types of textile dyes (600-800 Da), demonstrating a superior NF separation performance. Furthermore, the functionalized membranes display a distinct bactericidal activity with a pronounced reduction of 93.7% in the number of live *Escherichia coli* (*E. coli*) bacteria. This study highlights a fast, facile co-deposition strategy to assemble a multifunctional separation coating onto UF support, which renders a vast potential for textile wastewater treatment.

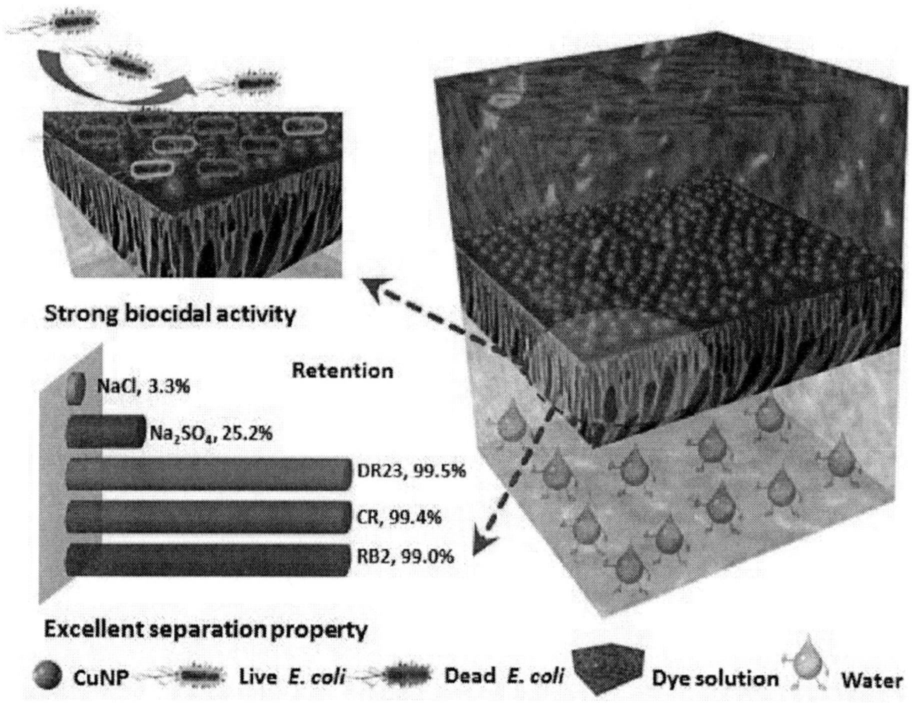

Reference:

[1] Zhu, J., Uliana, A., Wang, J., Yuan, S., Li, J., Tian, M., Simoens, K., Volodin, A., Lin, J., Bernaerts, K., Zhang, Y., & Van der Bruggen, B. (2016). Elevated salt transport of antimicrobial loose nanofiltration membranes enabled by copper nanoparticles via fast bioinspired deposition. *Journal of Materials Chemistry A*, *4*(34), 13211-13222.

AIChE
120 Wall Street, FL 23
New York, NY 10005-4020

ISBN 978-1-5108-3448-4

CURRAN ASSOCIATES INC.
proceedings.com